Polymer Science Library 1

CHEMORHEOLOGY
OF POLYMERS

Polymer Science Library

Edited by A.D. Jenkins
University of Sussex,
The School of Molecular Sciences,
Falmer, Brighton BN1 9QJ, England

1   *K. Murakami* and *K. Ono*, Chemorheology of Polymers

Polymer Science Library 1

CHEMORHEOLOGY OF POLYMERS

Kenkichi Murakami
and
Katsumichi Ono
Chemical Research Institute of Non-Aqueous Solutions,
Tohoku University, Sendai, Japan

ELSEVIER SCIENTIFIC PUBLISHING COMPANY
Amsterdam — Oxford — New York   1979

ELSEVIER SCIENTIFIC PUBLISHING COMPANY
335 Jan van Galenstraat
P.O. Box 211, 1000 AE Amsterdam, The Netherlands

Distributors for the United States and Canada:

ELSEVIER/NORTH-HOLLAND INC.
52, Vanderbilt Avenue
New York, N.Y. 10017

Library of Congress Cataloging in Publication Data

Murakami, Kenkichi, 1926–
  Chemorheology of polymers.

  (Library of polymer science ; 1)
  Includes index.
  1. Polymers and polymerization--Deterioration.
2. Rheology. I. Ono, Katsumichi, 1940-   joint
author. II. Title. III. Series.
QD381.8.M87   547'.84   79-19601
ISBN 0-444-41831-8

ISBN 0-444-41831-8 (Vol. 1)
ISBN 0-444-41832-6 (Series)

Printed in The Netherlands

PREFACE TO *POLYMER SCIENCE LIBRARY*

Polymer Science is one of the new disciplines to emerge in its own right as a subject appropriate to the needs of the second half of the twentieth century. Everyday life depends on the manufacture and processing of polymers to provide an amazing variety of plastics, fibres, films, paints and adhesives. Conventional materials, such as wood and metals, have given way to synthetic polymers for a multitude of applications, and the chemical industry as a whole is more concerned with polymers than with small molecules.

The life sciences too recognize that vital processes depend upon a number of naturally-occurring polymers, for example, proteins, carbohydrates and the nucleic acids, often in a way that requires the architecture of the polymer to be organized precisely, down to the most meticulous detail.

Even the most conventional aspects of polymer science are undergoing continuous development while new techniques, applications and concepts are making their appearance all the time. For this reason we hope in this collection of books to present accounts of the contemporary state of the art in the more established fields and to provide comprehensive guides to the more novel areas from all over the world. It is our hope to build up, in this way, an authoritative and topical library of polymer science, of value to everyone working in the field, whether from an academic or industrial point of view.

Brighton, England   A.D. Jenkins
June 1979

PREFACE

Many books dealing with degradation and stabilization of polymers
have already been published. The purpose of the present volume is
to give a condensed selection of the basic and fundamental knowledge
of polymer degradation and stabilization mechanisms, which other
books in this field have so far failed to present.

At sufficiently elevated temperatures, polymeric materials are subject to chemical reactions that affect their structural properties.
These structural changes are reflected in the visco-elastic behavior
of the polymeric materials. Such visco-elastic behavior, induced by
chemical reactions, may be termed chemorheology. Chemorheology is,
therefore, a new field of rheology that encompasses the physical and
chemical aspects involved in the elucidation of the mechanism, at a
molecular level, of structural changes of polymers during their
degradation.

We hope that the quantitative study of degradation and stabilization
presented here will assist further research and application in areas
where the lifetime of polymers may be controlled by means of additives,
chemical modifications and synthesis, and where this knowledge is
applied to thermal-, light-, oxidation-, ozon-, and chemical-resistant
polymers, flame retardant polymeric materials and disposable plastics.

During the writing of this volume, the authors are greatly indebted
to Drs. T. Kusano and S. Tamura, and other colleagues at our institute
for their helpful cooperation.

They are also indebted to Miss Kazue Shoji for the preparation of
the manuscript.

Sendai, Japan   Kenkichi Murakami
June 2, 1979   Katsumichi Ono

CONTENTS

X

CHAPTER 1

INTRODUCTION

The measurement of stress relaxation as a function of time in elastomers held at constant elongation has been shown by Tobolsky to be a useful tool in investigating chemical degradation (ref. 1). By varying the temperature and the time of the experiment, the viscoelastic behavior also can be studied. Such studies were termed chemorheology (refs. 2-9), which was gradually expanded and systematized by Tobolsky.

The phenomenon of chemical stress relaxation was discovered in an investigation of crosslinked (vulcanized) rubber (NR), chloroprene rubber (CR), butyl rubber (IIR), butadiene-styrene co-polymers (SBR), and butadiene-acrylonitrile co-polymers (NBR).

It was found that, in the temperature range 100 to 150°C, these vulcanized rubbers showed a fairly rapid decay to zero stress at constant extension. Since, in principle, a crosslinked rubber network in the rubbery range of behavior should show little stress relaxation, and certainly no decay to zero stress, the phenomenon was attributed to a chemical rupture of the rubber network. This rupture was specifically ascribed to the effect of molecular oxygen since under conditions of very low oxygen pressures (< 10 Pa) the stress-relaxation rate was markedly diminished. However, at moderately low oxygen pressures, the rate of chemical stress relaxation was the same as under atmospheric conditions. This result parallels the very long established fact that, in the liquid phase, the rate of reaction of hydrocarbons with oxygen is independent of oxygen pressure down to fairly low pressures.

It is essential to ensure that the rubber samples used in these studies are thin enough to permit steady-state conditions of oxygen diffusion; otherwise the rate of chemical stress relaxation may become diffusion-controlled.

In the original study of chemical stress relaxation, many other important facts were established. The relative stress-decay curves $f(t)/f(0)$, where $f(t)$ is the stress at time t, and $f(0)$ the initially measured stress, were found to be independent of the elongation at which the rubber was maintained up to values of at least 200% exten-

2

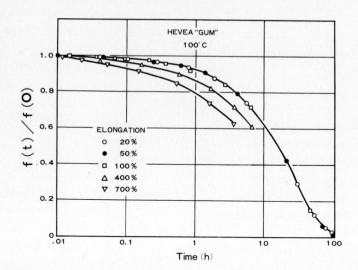

Fig. 1.1. Effect of elongation on chemical stress relaxation of sulfur-cured natural rubber at 100°C.

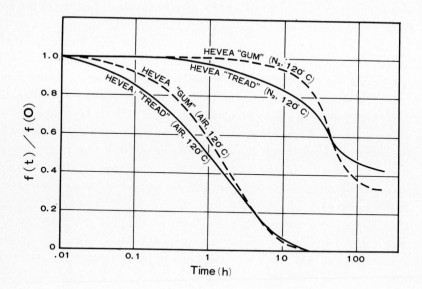

Fig. 1.2. Chemical stress-decay curves in air for sulfur-cured natural rubber. The "gum" contains no filler; the "tread" contains 33⅓% by weight of reinforcing carbon black. Also shown are the stress-decay curves of these same rubbers in highly purified nitrogen.

sion (extension ratio $\alpha = 3$). Some data of this type are shown for the natural-rubber vulcanizate in Fig.1.1. The corresponding stress-decay curves $f(t)/f(0)$ were found to be nearly independent of the presence or absence of carbon black or other fillers, as shown in Fig.1.2. Also shown in Fig.1.2 is an example of the marked reduction in rate of stress-relaxation which can be achieved by rigorous exclusion of oxygen. Complete exclusion is impossible, however, and it is of course impossible to remove peroxide groups along the chain simply by evacuation.

REFERENCES

1 A.V. Tobolsky, I.B. Prettyman, and J.H. Dillon, J. Appl. Phys., 15(1944)324, 280; Rubber Chem. Technol., 17(1944)551.
2 R.D. Andrews, A.V. Tobolsky and E.E. Hauson, J. Appl. Phys., 17(1946)352.
3 R.B. Beevers, J. Colloid Sci., 19(1964)40.
4 J. Scanlan, Trans. Faraday Soc., 57(1961)839.
5 L.E. St. Pierre and A.M. Bueche, J. Phys. Chem., 63(1959)1338.
6 J.R. Dunn, J. Scanlan and W.F. Watson, Trans. Faraday Soc., 27(1958)730.
7 J.P. Berry, Trans. Inst. Rubber Ind., 32(1956)224.
8 R.C. Osthoff, A.M. Bueche and W.T. Grubb, J. Am. Chem. Soc., 76(1954)4659.
9 A.V. Tobolsky and A. Mercurio, J. Polym. Sci., 36(1959)467.

CHAPTER 2

SINGLE DEGRADATION OF CROSSLINKED POLYMERS

I SCISSION ALONG MAIN CHAINS

1 The Effect of a Random Distribution of Chain-Length on Scission along Main Chains

In the case of crosslinked polymers, their structures are divided into two parts. One is a uniform network as shown in Fig.2.1(a), and the other is one having a random distribution of chain-length as shown in Fig. 2.1(b).

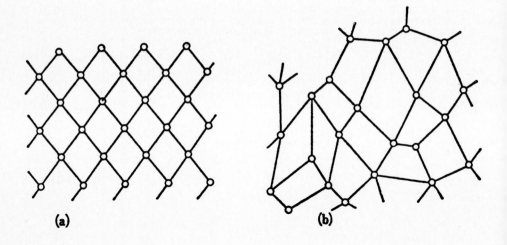

(a)   (b)

Fig. 2.1. Distribution of network chain-lengths. (a) Uniform chain-lengths.
(b) Random chain-lengths.

Bueche (ref. 1), and Berry and Watson (ref. 2) derived the equation expressing the relation between $q_m(t)$, which is the total number of scission along main chains per cubic centimeter, and relative stress $f(t)/f(0)$, assuming a random distribution of chain-length in a network. It is well known that the expression for the tensile stress $f(0)$ to deform a rubber to an extension ratio α based on the statistical theory of rubber-like elasticity is given by (ref. 3):

$$f(0) = n(0)RT(\alpha - \alpha^{-2}) = N(0)kT(\alpha - \alpha^{-2}) \qquad (2.1)$$

Here, $n(0)$ and $N(0)$ are the moles and the number of network chains per cubic centimeter of rubber, respectively. R, k and T are the gas constant, Boltzmann constant and absolute temperature, respectively.

Suppose that, at time t, there have been $q_m(t)$ cleavages of network chains per cubic centimeter of rubber, and that there are $n(t)$ moles of network chains per cubic centimeter (or $N(t)$) that are still supporting the stress $f(t)$. Then,

$$f(t) = n(t)RT(\alpha - \alpha^{-2}) = N(t)kT(\alpha - \alpha^{-2}) \qquad (2.2)$$

From eqns (2.1) and (2.2), one obtains,

$$\frac{f(t)}{f(0)} = \frac{n(t)}{n(0)} = \frac{N(t)}{N(0)} \qquad (2.3)$$

(A) The Equation of Berry and Watson (ref. 4)

The initial number of elastically active network chains of length x in a random distribution of such chains is given by

$$N_x(0) = N(0) \cdot p \cdot (1 - p)^{x-1} \qquad (2.4)$$

where $N(0)$ is the total number of chains and p is the reciprocal of average chain-length. If the probability, that a repeating unit has undergone a scission in time t, is β, the number of uncut chains of length x is

$$N_x(t) = N(0) \cdot (1 - \beta)^x \qquad (2.5)$$

provided that x is large enough for the chain ends to be neglected; the crosslinked units on both ends of a chain do not contribute appreciably to the length of the chain. If we further assume that the probability β is the same for all repeating units of chains of different length, then eqns (2.4) and (2.5) can be combined and summed to give

$$\frac{N(t)}{N(0)} = \frac{p(1 - \beta)}{p + \beta(1 - p)} \qquad (2.6)$$

Here,

$$N(t) = \sum_{x=1}^{\infty} N_x(t) \qquad (2.6)$$

This is the equation of Berry and Watson.

Assuming that M_o is the number of repeating units per cubic centimeter (unit volume), the relation of $M_o = x \cdot n(0)$ exists and β is expressed by

$$\beta = \frac{q_m(t)}{M_o} \qquad (2.7)$$

Substitution of eqn (2.7) into eqn (2.6) gives:

$$\frac{q_m(t)}{M_o} = \frac{1 - \dfrac{N(t)}{N(0)}}{1 + \dfrac{N(t)}{N(0)}(\dfrac{1}{p} - 1)} \qquad (2.8)$$

(B) The Equation of Yu (ref. 5)

Yu derived his equation from a different point of view. Let $q_m(t)$ be the total number of random scissions that have occurred at time t per unit volume of a network. Then, the number of scissions occurring in the chains of length x would be

$$q_{m,x}(t) = q_m(t) \cdot \frac{x N_x(0)}{\sum\limits_{x=1}^{\infty} x \cdot N_x(0)} \qquad (2.9)$$

This simply apportions the scissions according to the number fractions of repeating units without specifying the probability of scission of the individual unit. Thus, the number of uncut chains of length x after time t is

$$N_x(t) = N_x(0) \left\{ 1 - \frac{1}{N_x(0)} \right\}^{q_{m,x}(t)} \qquad (2.10)$$

since $1/N_x(0)$ is the probability that one of these chains undergoes cleavage at each scission. Implicit in eqn (2.10) is the assumption that the probability $1/N_x(0)$ does not change while $q_{m,x}(t)$ scissions are taking place. This assumption is, strictly speaking, effective for a scission mechanism of the interchange type. Substituting eqn (2.4) into (2.10) and summing over all chains, one arrives eqn (2.11), which is the equation of Yu.

$$\frac{N(t)}{N(0)} = \frac{p}{\exp\left\{\dfrac{pq_m(t)}{N(0)}\right\} + p - 1}$$

(2.11)

By substitution of $M_o = x \cdot N(0)$ described above, eqn (2.11) is modified to

$$\frac{q_m(t)}{M_o} = \ln\left\{ p\left(\frac{N(0)}{N(t)} - 1\right) + 1 \right\}$$

(2.12)

(C) The Equation of Tobolsky (ref. 6)

For the case of uniform network chain-length, eqns (2.13) and (2.14) are obtained.

$$N_x(t) = N(t)$$

(2.13)

$$N_x(0) = N(0)$$

(2.14)

Substituting eqns (2.13), (2.14) and (2.7) into eqn (2.5), one obtains

$$\frac{N(t)}{N(0)} = \left\{ 1 - \frac{q_m(t)}{M_o} \right\}^x = \left\{ 1 - \frac{q_m(t)}{x \cdot N(0)} \right\}^x$$

(2.15)

When the average chain-length x between two crosslinkages is large enough, eqn (2.15) is approximated by

$$\frac{N(t)}{N(0)} = \left\{ 1 - \frac{q_m(t)}{x \, N(0)} \right\}^x \approx e^{-\dfrac{q_m(t)}{N(0)}}$$

(2.16)

Therefore,

$$\frac{q_m(t)}{M_o} = p \, \ln\frac{N(0)}{N(t)}$$

(2.17)

(D) The Relationships between the above Theoretical Equations

From inspection of eqns (2.8) and (2.12), it is evident that the two reduce to the same equation at the limit of small p as shown in the following equations.

In the case of $p \ll 1$,
from the equation of Berry and Watson,

$$\frac{q_m(t)}{M_o} = \frac{1-\dfrac{N(t)}{N(0)}}{1 + \dfrac{N(t)}{N(0)}(\dfrac{1}{p} - 1)} \approx \frac{1 - \dfrac{N(t)}{N(0)}}{\dfrac{1}{p} \cdot \dfrac{N(t)}{N(0)}} = p\left\{\frac{N(0)}{N(t)} - 1\right\} \quad \text{and} \quad (2.18)$$

from the equation of Yu,

$$\frac{q_m(t)}{M_o} = \ln\left\{p(\frac{N(0)}{N(t)} - 1) + 1\right\} \approx p\left\{\frac{N(0)}{N(t)} - 1\right\} \quad (2.19)$$

When the probability $q_m(t)/M_o$, that a repeating unit has undergone a scission in time t is plotted against the fraction of network chains $N(t)/N(0)$, at time t, Fig. 2.2 is obtained.

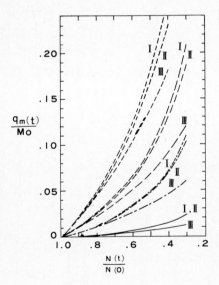

Fig. 2.2. Random chain-scission of network. $q_m(t)/M_o$, cuts per repeating unit as a function of the fraction of network chains at time t, $N(t)/N(0)$. (---) $1/p=5$; (— — —) $1/p=10$; (— - — -) $1/p=20$; (———) $1/p=100$, I(eq 2.8), II(eq 2.12), III(eq 2.17).

We note from Fig. 2.2 that, as $1/p$ approaches 100 or more (i.e., approaches the actual crosslinking for rubber vulcanizates) the more meaningful is the approximation of uniform chain-length in estimating the number of scissions from the relative stress, if the mechanism of random scission is operating. On the other hand, as p approaches unity (i.e., the higher the crosslinking) the less serious is the approximation of uniform chain-length implicit in the equation of Tobolsky. In other words, as the average chain-length becomes smaller, the differences between the three curves corresponding to eqns (2.8), (2.12) and (2.17) are appreciable.

So, it is clear from Fig. 2.2 that eqn (2.17) for uniform chain-length network is useful for all actual rubber vulcanizates, even if they have a random distribution of chain-length. Recently, a method that the approximate chain-length distribution of crosslinked polymers can be extracted from stress-relaxation measurements was presented by Curro (ref. 7).

From eqns (2.26) and (2.30) in the following section,

$$N_x(t) = N_x(0) \cdot e^{-x \cdot k_1 \cdot t} \tag{2.20}$$

The total number of elastically effective chains in this system $N(t)$ is just the sum over all chain-lengths.

$$N(t) = \sum_{x=1}^{\infty} N_x(t) = \sum_{x=1}^{\infty} N_x(0) \cdot e^{-x \cdot k_1 \cdot t} \tag{2.21}$$

If both sides of this equation are divided by $N(0)$, the initial number of chains, we obtain the desired relation:

$$\frac{N(t)}{N(0)} = \sum_x P_x \cdot e^{-x \cdot k_1 \cdot t} \tag{2.22}$$

where P_x is the initial normalized fraction of chains having length n. In combination with eqns (2.3) and (2.22) becomes

$$F(k_1 t) = \frac{f(t)}{f(0)} = \sum_x P_x \cdot e^{-x \cdot k_1 \cdot t} \tag{2.23}$$

and eqn (2.23) is modified to the more natural form,

$$F(k_1 t) = \int_o^\infty P(x) \cdot e^{-k_1 \cdot tx} \, dx \qquad\qquad (2.24)$$

The force on the network is the Laplace transform of the distribution function $P(x)$. The low-order approximations for $P(x)$ in eqn (2.24) work well as long as the function $P(x)$ does not have sharp changes.

In order to demonstrate the value of this method, stress-relaxation data from Tobolsky et al. were used (ref. 8). These data, taken at 130°C on radiation-crosslinked natural rubber, appear in Fig. 2.3. Radiation-crosslinked natural rubber at 130°C fulfils the requirements

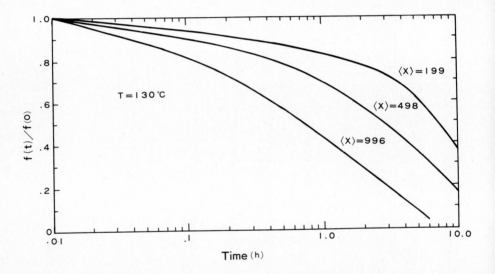

Fig. 2.3. Stress-relaxation data of Tobolsky et al. on radiation-cured natural rubber.

that we have random scission along the main chain and that chemical reaction should provide the major relaxation mechanisms at this temperature. The data were numerically differentiated at various points and the first and second derivative approximation method was used to obtain the chain-length distributions shown in Fig. 2.4. If the radiation crosslinking process were completely random, $P(x)$ would have the form:

$$P(x) = \frac{(1 - x^{-1})^{-1}}{x} \qquad\qquad (2.25)$$

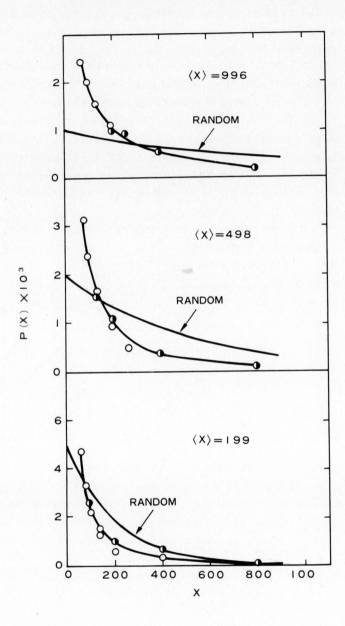

Fig. 2.4. Calculated chain-length distributions. Key: (O), first derivative approximation, (◑), second derivative approximation. The random crosslinking curve was calculated. k was found to be 0.0025l/h.

These curves are also shown in Fig. 2.4.

It can be observed from Fig. 2.4 that the results obtained for P(x) are weighted more toward short chain-length than the random cross-linking process predicted by eqn (2.25). A possible explanation for this is that the bimolecular crosslinking process would be expected to occur with a higher probability in the neighbourhood of an already existing crosslink, thereby favoring short chain-lengths.

In summary, the chain-length distribution of elastically-effective chains in elastomers can be estimated from high-temperature stress-relaxation experiments, provided the sample is known to undergo random scission along main chains. This can be accomplished by numerical differentiation of the data and use of the approximation methods.

2 Basic Mechanism of Scission along Main Chains

When scission along main chains occurs for an uniform chain-length network, eqn (2.26) is obtained from eqn (2.16) suggested by Tobolsky and eqn (2.3).

$$\frac{f(t)}{f(0)} = \frac{N(t)}{N(0)} = e^{-\frac{q_m(t)}{N(0)}} \qquad (2.26)$$

From eqn (2.26), one obtains

$$q_m(t) = -N(0) \ln \frac{f(t)}{f(0)} \qquad (2.27)$$

which is an expression for estimating $q_m(t)$ from the relative stress $f(t)/f(0)$. Another way of deriving eqn (2.27) is as follows.

Suppose a number of additional scissions, $\Delta q_m(t)$ (i.e., 8 in Fig. 2.5) occurs during the interval Δt and the number of scissions which occurs in the region of uncut chains (the solid lines in Fig. 2.5) is $\Delta N(t)$ (i.e., 2 in Fig. 2.5), the value of $\Delta N(t)/\Delta q_m(t)$ (i.e., 1/4 in Fig. 2.5) will be indicated by the ratio of nN(t) to nN(0) (n is a number of repeated units between two crosslinkages), or be given by

$$- \frac{\Delta N(t)}{\Delta q_m(t)} = \frac{N(t)}{N(0)} \qquad (2.28)$$

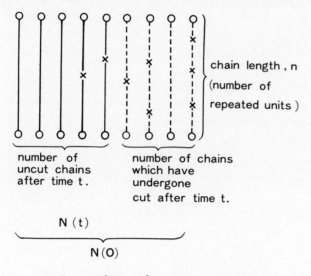

chain length , n
(number of
repeated units)

number of
uncut chains
after time t.

number of chains
which have
undergone
cut after time t.

N (t)

N (O)

total number of networK chains

Fig. 2.5. Mechanism of network chain-scissions x: scission point.
$\Delta q_m(t)$ is the total number of x points.

The cutting point x is included in each uncut chain only once
because, although such chains may ultimately suffer several scissions,
after the first cut they no longer count as uncut chains.

The more general form of eqn (2.28) is given by

$$- \frac{dN(t)}{dq_m(t)} = \frac{N(t)}{N(0)} \tag{2.29}$$

which can be integrated to give

$$q_m(t) = - N(0) \ln \frac{N(t)}{N(0)} = - N(0) \ln \frac{f(t)}{f(0)}$$

It will be assumed in many cases that the rate of scission is
approximately constant, as is apparent from the experimental results
(ref. 9). Thus

$$q_m(t) \approx q_0 t \tag{2.30}$$

Here, q_o is the number of scissions ocurring along the main chain in unit time. Substituting eqn (2.30) into (2.26), one obtains

$$\frac{f(t)}{f(0)} \approx e^{-k_1 t} \tag{2.31}$$

Accordingly, chemical relaxation will be approximated by the Maxwellian decay curve of eqn (2.31), when scission along main chains occurs in the crosslinked polymers. Maxwellian decay curves are shown in Fig. 2.6.

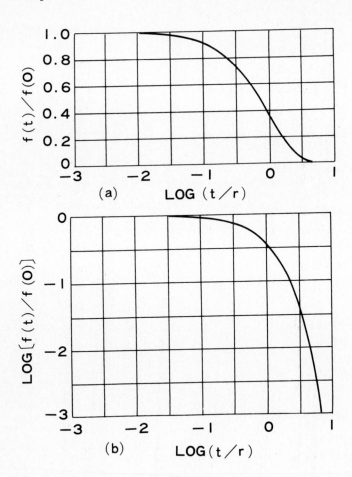

Fig. 2.6. Maxwellian decay of stress at constant extension.

The characteristic behavior of Maxwellian decay is that, when $f(t)/f(0)$ is plotted against log t, the decay occurs over two cycles of logarithmic time, as shown in Fig. 2.6. This is very definitely true for natural rubber, as can be seen in Fig. 2.7. For the other hydrocarbon rubbers (GR-S, Neoprene, Butyl, GR-N, etc.), chemical stress decay, though not exactly Maxwellian, was also found to occur in two to three cycles of logarithmic time, as shown in Fig. 2.7.

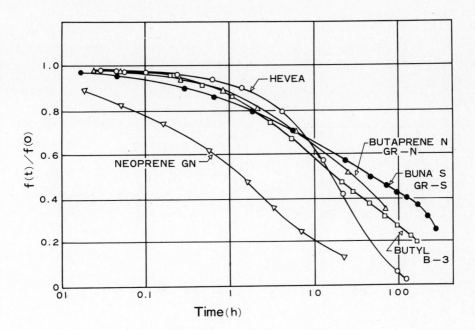

Fig. 2.7. Chemical stress-decay curves for several synthetic rubber vulcanizates of hydrocarbon-type structure at 100°C.

For all these rubbers, it was possible to describe the effect of temperature on chemical stress decay as follows,

$$\frac{f(t)}{f(0)} = \phi \; (k',t) \qquad (2.32)$$

where k' is a function of T alone. This, of course, means that, if $f(t)/f(0)$ for a given rubber is plotted against log t, the decay curves at various temperatures can be superposed by horizontal translation along the logarithmic time axis. The functional dependence of k' on temperature was found to be expressed by the Arrhenius equation.

$$k' = A \cdot \exp(- E_a/RT) \qquad (2.33)$$

For the rubbers shown in Fig. 2.7, E_a was found (Ref. 10) to be abc 127 kJ mol^{-1}. For the particular natural rubber vulcanizate used in this study, it was found that

$$k' = 8.1 \times 10^{12} \exp(- 127/RT)$$

The predicted stress-decay curves based on the above equation and eqn (2.31) at 100, 110, 120 and 130°C are shown in comparison with the experimental data in Fig. 2.8 plotted in the form log f(t)/f(0) vs. t.

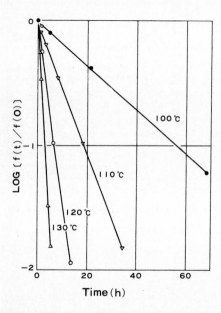

Fig. 2.8· Effect of temperature on chemical stress-decay for a natural-rubber vulcanizate. Lines are computed from eqns (2.31) and (2.33). Points are experimenta data.

Chemical stress relaxation therefore becomes a useful physical method for measuring chemical-bond cleavage in rubber networks. The rate of this cleavage reaction was considered to be completely independent of the forces acting on the rubber network and of the elongation of the network (up to very high elongations).

Assuming a random distribution of chain-length network, the Berry-Watson treatment yields the following simple result (Ref. 4), if the average length x is large:

$$q_m(t) = N(0) \left\{ \frac{f(0)}{f(t)} - 1 \right\} \tag{2.34}$$

Another approach for relating $q_m(t)$ to $f(t)/f(0)$ and $N(0)$ has been
used by Scanlan (ref. 11). He challenges the premise that $N(t)/N(0)$
is equal to the probability that a given chain has not been cut. His
treatment hinges on the idea that only a finite number of cleavages
are necessary before an insoluble network (gel) is reduced to a
soluble polymer (sol) incapable of supporting a force. His final result
is expressed as follows:

$$\frac{f(t)}{f(0)} = (1 - S_1)^3 (1 + S_1) \tag{2.35}$$

where

$$S_1 = -\frac{1}{2} + \left\{ \frac{1}{4} + \frac{q_m(t)}{N(0)} \right\}^{\frac{1}{2}}$$

This result gives $f(t)/f(0)$ equal to zero when $q_m(t)/N(0)$ is equal to
2.

The benzoyl peroxide-initiated oxidative scission of crosslinked
poly(ethyl acrylate) was studied by the methods of chemical stress
relaxation and oxygen absorption by Tobolsky et al. (ref. 12), and
eqn (2.36) was derived by them, assuming the decay of initiator to
follow a first-order rate law:

$$q_m(t) = 2e_S I_o (1 - e^{k_d t}) \tag{2.36}$$

where I_o is the initiator concentration, k_d is the specific rate
constant for homolytic cleavage of benzoyl peroxide, and e_S is the
efficiency of scission, the value of which is 0.23 for poly(ethyl
acrylate). The value of k_d for benzoyl peroxide at 85°C is 6.1 x 10^{-5}
s^{-1} (ref. 73).

If the value of $q_m(t)$ is plotted against time for the various
theoretical relations, eqns (2.27), (2.34), and (2.35), Figures 2.9,
2.10 and 2.11 are obtained.

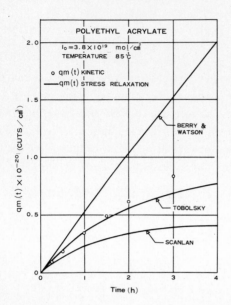

Fig. 2.9. $q_m(t)$ vs. time for polyethyl acrylate vulcanizates using I_o=2.5 x 10^{19} mol cm^{-3}.

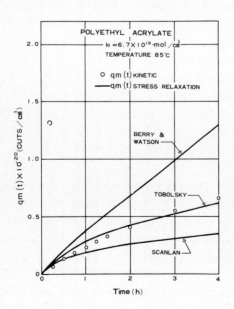

Fig. 2.10. $q_m(t)$ vs. time for polyethyl acrylate vulcanizates using I_o=3.8 x 10^{19} mol cm^{-3}.

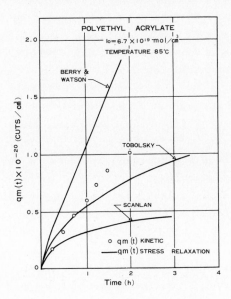

Fig. 2.11· $q_m(t)$ vs. time for polyethyl acrylate vulcanizates using I_O=6.7 x 10^{19} mol cm^{-3}.

If the kinetic prediction of scission by eqn (2.36) is valid, this estimate of the number of scissions tends to follow the number calculated by the Tobolsky eqn (2.27) over the entire range of I_O; the reason can be considered as follows.

The random distribution weights (refs. 13,14) the network chains of small size, very disproportionately, and small network chains are to some extent prevented from forming interpenetration by other network chains. Furthermore, very short network chains would not be elastic elements of the rubber network, but might act instead as long crosslinks.

Recently Shaw (ref. 15) compared the theoretical relations between f(t)/f(0) and time based upon Tobolsky's, Flory's and Scanlan's equations with his experimental results obtained by using the artificial simpler models of polymer networks proposed by him for the case of random scission of main chains and of scission of crosslinkages.

3 Maxwellian Decay and the Modified Chemical Relaxation Curve (ref. 16)

As described above, chemical stress relaxation was found to be approximated by Maxwellian decay for hydrocarbon rubber vulcanizates. More careful experiments, carried out later, indicated that the

chemical relaxation was not necessarily expressed by Maxwellian decay
for rubber vulcanizates. The degree of deviation from linear behavior
when log f(t)/f(0) is plotted against time is different, depending on
the kind of samples and other conditions. Once deviated, however, the
the tendency of its observed exponential decay was found to be shown
downward from the straight lines.

This tendency is not observed for sulfur-cured natural rubber and
gutta-percha (*trans*-1,4-polyisoprene) for which the stress decay
curves are generally Maxwellian. As for peroxide-cured rubbers,
however, the Maxwellian decay curves are followed by downward deviati
in the later stages. If the rubber sample is purified by reprecipitation before curing, the stress decay is extremely fast and deviation
from the Maxwellian curve is noticeable, even in the initial stages.
An example of such behavior will be shown below for gutta-percha in
some detail.

The Maxwellian plots of stress-decay curves are shown in Fig. 2.12
in which the deviation from straight lines is clear.

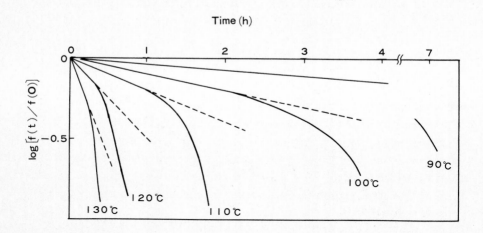

Fig. 2.12.log f(t)/f(0) vs. time t curves at various temperatures in air for
1,4-polyisoprene.

The gutta-percha used here was milled with dicumyl peroxide (DCP),
then the sheets of specimen were prepared by compression molding at
150°C and were extracted with hot acetone for 48 h (ref. 17).

As described above, appreciable deviation from Maxwellian decay
is observed for non-purified gutta-percha, and such deviation becomes
more noticeable after purification. The measurements were carried out

between 90 and 130°C in air.

The logarithm of the relative stress $f(t)/f(0)$ was plotted against time t, as shown in Fig. 2.12. An induction period was observed at every temperature. The higher rate of the chain scission in the later stages suggests that the degradation process of DCP-cured gutta-percha is autocatalytic, as in the case of DCP cured natural rubber. We propose an empirical equation of the following type for the number of main-chain scissions $q_m(t)$ at time t.

$$q_m(t) = at + b \cdot \exp(ct) \qquad (2.37)$$

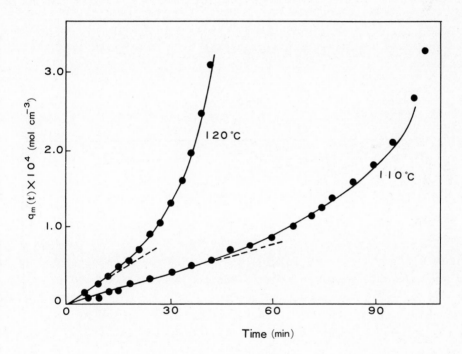

Fig. 2.13. $q_m(t)$ vs. time t curves at 110°C, and 120°C for *trans*-1,4-polyisoprene, ● : exptl., —— : calc. from eqn (2.37).

The parameter a in eqn (2.37) was determined from the initial slope of $q_m(t)$ vs. t curves, while parameters b and c were estimated by

plotting the logarithm of $\{q_m(t)-at\}$ vs. t. The sets of these parameters are listed in Table 2.1. Using these values, $q_m(t)$ was

TABLE 2.1

Degradation parameters in eqn (2.37)

Temp. in °C	$10^7 \cdot a$ mol cm^{-3} min^{-1}	b mol cm^{-3}	c min^{-1}
90	2.35	8.71×10^{-9}	0.0215
100	6.68	4.68×10^{-8}	0.0334
110	13.4	3.39×10^{-7}	0.057
120	27.6	1.31×10^{-6}	0.116
130	56.9	4.67×10^{-6}	0.159

calculated from eqn (2.37) and was compared with the experimental
points as shown in Fig. 2.13. As seen from this figure, the rate of
network scission in gutta-percha is well approximated by eqn (2.37).

According to conventional mechanism of autoxidation (ref. 18),
we have:

$$RH + O_2 \xrightarrow{w_i} R\cdot \tag{i}$$

$$R\cdot + O_2 \xrightarrow{k_2} RO_2^\cdot \tag{ii}$$

$$RO_2^\cdot + RH \xrightarrow{k_3} ROOH + R\cdot \tag{iii}$$

$$ROOH + RH \xrightarrow{k_4} \delta RO_2^\cdot \tag{iv}$$

where w_i is the rate of initiation and δ is the average number of the
peroxy radicals produced in the reaction between ROOH and RH.

In the initial stage of oxidation, $q_m(t)$ increases almost linearly
with time. Such behavior is characteristic for inhibited autoxidation
(ref. 19) and suggests that naturally-occurring impurities of gutta-
percha, which could not be extracted by hot acetone, are playing the
role of antioxidants. To confirm this hypothesis, the stress relaxati
of gutta-percha (purified by reprecipitation before curing) was exami
The rate of stress decay was extremely fast and no induction period
was apparent.

If the inactivation of peroxy radical by impurities (AH) is the
only important termination reaction, then

$$RO_2^\bullet + AH \xrightarrow{\quad k_5 \quad} \text{inactive products} \qquad\qquad (v)$$

It was also assumed that the rate of scission is proportional to peroxy radical concentration.

$$\frac{dq_m(t)}{dt} = k \left[RO_2^\bullet\right] \qquad\qquad (2.38)$$

Under the conditions of steady-state concentrations of alkyl and peroxy radicals,

$$\left[RO_2^\bullet\right] = \frac{1}{k_5[AH]_o} \left(w_i + k_4\delta[RH]\cdot[ROOH]\right) \qquad\qquad (2.39)$$

where $[AH]_o$ is the initial concentration of impurity. If the hydroperoxide concentration is negligible in the early stage, we obtain

$$\left[RO_2^\bullet\right] = \frac{w_i}{k_5[AH]_o} \qquad\qquad (2.40)$$

From eqns (2.38) and (2.40),

$$\frac{dq_m(t)}{dt} = \frac{kw_i}{k_5[AH]_o} \qquad\qquad (2.41)$$

Evidently, the value of $dq_m(t)/dt$ in eqn (2.41) corresponds to the constant a in eqn (2.37), i.e., to the slope of the linear part observed in the $q_m(t)$ vs. t curve. Furthermore, eqn (2.41) leads to eqn (2.30), which applies in many cases of rubber vulcanizates, the chemical relaxation of which is expressed by Maxwellian decay.

In the later stage, on the other hand, the oxidation conditions can not be regarded as stationary. Under steady-state conditions for alkyl radicals, a pair of differential equations will be obtained, thus:

$$\frac{d\left[RO_2^\bullet\right]}{dt} = w_i + k_4\delta[ROOH] \qquad\qquad (2.42)$$

$$\frac{d[\text{ROOH}]}{dt} = k_2[\text{RH}][\text{RO}_2^{\cdot}] - k_4\delta[\text{ROOH}] \qquad (2.43)$$

Here, we omitted eqn (v) from the assumption that the concentratic of AH is small. The solution of eqn (2.42) has the following form:

$$[\text{RO}_2] = a_1 e^{\lambda_1 t} + a_2 e^{\lambda_2 t} + b_1 \qquad (2.44)$$

where λ_2 is negative and other parameters are positive. At large t, only the first term of eqn (2.44) is important. This term corresponds to the second term of eqn (2.37). Therefore, eqn (2.37) satisfies the critical conditions given by eqns (2.41) and (2.44).

Of course, we cannot say anything about the mechanism of chain scission from the analysis of stress-relaxation curves; the treatment above shows that the oxidative degradation process of DCP-cured gutta-percha is well explained by the free radical mechanism of chain branching. The rate of stress decay for DCP-cured gutta-percha seems to be much faster than that of DCP-cured natural rubber. In general, the rate of stress decay for purified samples appears to be faster tha that of unpurified ones, which means that the stress-decay curves for purified samples show greater deviation from Maxwellian decay than those for unpurified ones, as is shown in Fig. 2.14.

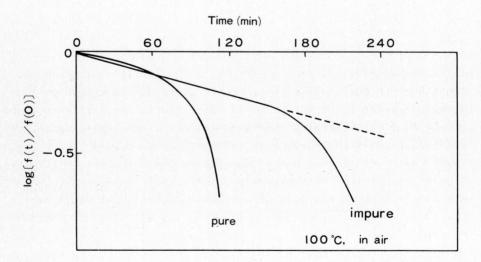

Fig. 2.14. log f(t)/f(0) vs. time t curves for unpurified and purified *trans*-1,4-polyisoprene.

Non-Maxwellian decay is represented by eqn (2.45) which was derived from eqns (2.26) and (2.37),

$$\ln \frac{f(t)}{f(0)} = -k_1 t - b \cdot \exp(ct) \qquad\qquad (2.45)$$

while Maxwellian decay is indicated by eqn (2.46) from eqn (2.31).

$$\ln \frac{f(t)}{f(0)} = -k_1 t \qquad\qquad (2.46)$$

4 Diffusion Effects on Chemical Stress Relaxation

The rate of oxidation in solid polymers is generally determined by two different processes. The first is the rate of diffusion of oxygen molecules into solid polymers, and the second is the rate of reaction of oxygen with the polymer chains. In the liquid phase oxidation, the first process is so fast that the overall rate of oxidation is scarcely controlled by the diffusion of oxygen. On the other hand, the oxidation of glassy or crystalline polymers is almost completely controlled by the rate of diffusion of oxygen on account of the small value of the diffusion coefficient for oxygen in the solid phase.

For rubbers or rubber vulcanizates, the values of diffusion coefficients are intermediate between those of liquids and glassy solids, and therefore the degree of "diffusion control" in the oxidation of rubbers depends on sample dimensions and temperature. Since the activation energy for the oxidation of rubbers is much higher than that for diffusion, the effect of diffusion becomes prominent at higher temperatures.

The importance of diffusion effects in the degradation of rubber vulcanizates was first pointed (ref. 20) out by Shelton et al. They investigated the rate of O_2 absorption of various rubber vulcanizates with different film thickness and found that the rate of O_2 uptake per unit volume of the film is substantially constant if the film thickness is less than a critical value. The effect of temperature on this critical value was also investigated. Their results are shown in Table 2.2. Below the critical sample thickness, the oxidation may be regarded as being free from diffusion control.

The effect of sample thickness on chemical stress relaxation was

TABLE 2.2

Critical sample thickness shown to be free from diffusion control (ref. 20)

Vulcanizate	Temp. (°C)	Sample thickness (cm)	Vulcanizate	Temp. (°C)	Sample thickness (cm)
SBR-Black	110	0.075	NR-Gum	60	0.075
	120	0.040		80	0.040
	130	0.040		100	0.020
	150	0.020			
SBR-Gum	110	0.075	Butaprene-Black	90	0.040
	120	0.075		100	0.040
	130	0.020		130	0.015
NR-Black with antioxidant	80	0.075	Butaprene-Gum	90	0.040
	100	0.075		100	0.020
	110	0.040		130	0.10
NR-Black without antioxidant	60	0.075	Neoprene-Black	90	0.040
	80	0.040		100	0.040
	90	0.040		110	0.020
	100	0.040			
			Neoprene-Gum	90	0.045
				100	0.045
				110	0.025

investigated by Berry (ref. 21). He concluded that the stress relaxat
curves are unaffected by sample thickness from 0.1 to 0.5 mm at 80 to
120°C.

Although these results served to establish the correct experimenta
conditions for chemorheology, the investigations were only qualitativ
ones, and none of the conclusions was concerned with the relative
contribution of diffusion and chemical reaction.

Many restrictions exist in the quantitative estimation of diffusio
effects on the oxidation of rubbers. In general, the oxidation of
hydrocarbon polymers is a complex reaction, including many elementary
reactions. The oxidation process in these polymers is autocatalytic
in nature, and the steady state for the reaction itself is not
necessarily reached. The analysis of such processes is hardly possibl
in practice, since the solution of the diffusion equation cannot be
obtained analytically. Therefore, the investigation is inevitably
confined to the initial stage of oxidation, which may often be
approximated to a steady reaction. Fortunately, rubbers are regarded
as homogeneous polymer liquids in the temperature range usually
employed for the studies of oxidation. The fraction of monomer units
taking part in the formation of crosslinking points is so small that
the whole system can be regarded as a homogeneous phase. Hence, the
rate coefficient of oxidation and the diffusion coefficient of oxygen

may be regarded as having constant values throughout the system.

The remainder of the present section is concerned with the quantitative estimation of diffusion effects on the chemical stress relaxation (refs. 22,23). In order to avoid complication, the following assumptions were made to formulate the relationship between diffusion and the rate of chain scission.

1) The diffusion coefficient D for the diffusion of O_2 into the rubber film is independent of O_2 concentration. It is generally recognized that the diffusion coefficients of simple gases are virtually independent of concentration.

2) The rate of O_2 consumption within the film obeys first order kinetics with respect to O_2 concentration C.

3) Scission occurs randomly along the main chain. This means that the concentration of available sites for scission is kept virtually constant throughout the reaction.

4) At a constant temperature, the rate of chain scission q is proportional to the rate of O_2 consumption k, that is,

$$\varepsilon = k/q \qquad (2.47)$$

where ε is scission efficiency.

Consider an infinite film of thickness 2ℓ. Let the film surfaces $x = \pm \ell$ be maintained at a constant O_2 concentration C_s with the initial concentration zero throughout the film, as illustrated in Fig. 2.15.

Edge effects may be safely neglected, if rubber strips with sufficient width are used in the chemical stress-relaxation measurements. When the diffusion of O_2 is coupled with first-order O_2 consumption by an irreversible reaction, the one-dimensional diffusion equation in the x direction is of the form (ref. 24),

$$\frac{\partial C}{\partial t} = D \frac{\partial^2 C}{\partial x^2} - kC = D \frac{\partial^2 C}{\partial x^2} - \varepsilon q C \qquad (2.48)$$

where C is the concentration of O_2 in the film, D is the diffusion coefficient, and k is the first-order rate constant for O_2 consumption. In the steady state, the solution may be obtained merely by putting $\partial C/\partial t = 0$.

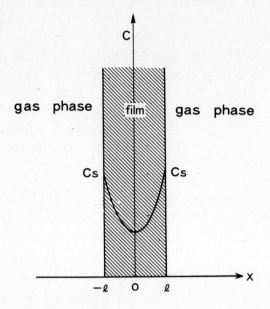

Fig. 2.15. Coordinates for a film of vulcanizate with thickness 2 ℓ.

$$D \frac{\partial^2 C}{\partial x^2} = kC \qquad (2.49)$$

The solution of eqn (2.49), satisfying the boundary conditions $C=C_s$ at $x=\ell$ and $\partial C/\partial x=0$ at $x=0$, is

$$C = C_s \cosh(\sqrt{k/D} \cdot x)/\cosh(\sqrt{k/D} \cdot \ell) \qquad (2.50)$$

Using Fick's first law of diffusion, we obtain an expression for the rate of O_2 uptake per unit area of the film.

$$F = 2D(\partial C/\partial x)_{x=\ell} = 2C_s \sqrt{Dk} \tanh(\sqrt{k/D} \cdot \ell) \qquad (2.51)$$

The rate of O_2 uptake per unit volume is given by,

$$F/2\ell = (C_s \sqrt{Dk}/\ell) \ \tanh(\ \sqrt{k/D}\cdot\ell) \qquad\qquad (2.52)$$

As shown previously, for the case of random scission along the main chain, the relationship between the relative stress $f(t)/f(0)$ and the rate of chain scission of a rubber network q_o is approximately given by eqn (2.31).

$$\frac{f(t)}{f(0)} = e^{-k_1 t} = e^{-q_o t/N(0)} \qquad\qquad (2.31)$$

When the scission reaction is diffusion-controlled, q_o becomes a function of x. Introducing the average rate of chain scission q_{av} such that,

$$q_{av} = \frac{1}{\ell} \int_o^\ell q_o(x)dx \qquad\qquad (2.53)$$

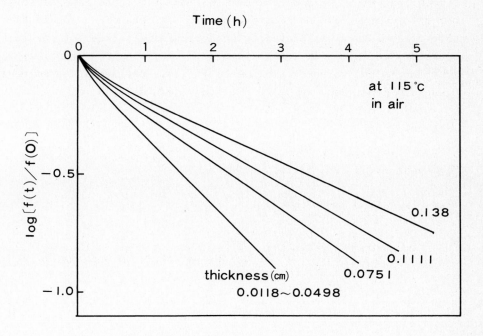

Fig. 2.16. $\log f(t)/f(0)$ vs. time t curves for various film thickness ($f(0)$: initial stress; $f(t)$: stress at time t).

then,

$$\log[f(t)/f(0)] = -q_{av}t/N(0) \qquad (2.54)$$

The values of q_{av} are easily determined from the plots of $\log[f(t)/f(0)]$ versus t for samples with various film thickness. An example for sulfur-cured synthetic *cis*-1,4-polyisoprene is shown in Fig. 2.16 (ref. 22). For all samples, stress-decay curves can be represented approximately by straight lines, except at the initial stage where faster stress decay was observed. The viscous flow of trapped entanglements and the exchange reaction at the crosslinks may both be responsible for the faster stress decay, but our interest is only in the later stages.

   Eqn (2.54) rests on the hypothesis that the concentration of the available sites for scission remains constant throughout the measurements. As described previously, this is one of the prerequisites for the applicability of eqn (2.48). If the scission occurs randomly concomitant with the chemical combination of O_2, the rate of scission per unit volume must be proportional to the rate of O_2 uptake per unit volume. This was proved to be approximately true for natural rubber vulcanizates by Tobolsky et al. (ref. 25). The proportionality constant is the scission efficiency ε, i.e., the number of molecules O_2 absorbed per cut, which have already been defined.

$$F/2\ell = \varepsilon q_{av} \qquad (2.55)$$

From eqns (2.52) and (2.55), we obtain,

$$q_{av}\ell = (C_s \sqrt{Dk}/\varepsilon)\ \tanh(\ \sqrt{k/D}\cdot\ell) \qquad (2.56)$$

If a dimensionless parameter $\sqrt{k/D}\cdot\ell$ takes a small value, eqn (2.56) can be reduced to,

$$q_{av} \simeq C_s k/\varepsilon \qquad (2.57)$$

This means that the rate of scission is almost independent of the diffusion coefficient in the case of a thin film. On the other hand, if $\sqrt{k/D}\cdot\ell$ takes a large value,

$$q_{av}\ell \simeq C_s \sqrt{Dk}/\varepsilon \qquad\qquad (2.58)$$

eqn (2.58) shows that the product $q_{av}\ell$ approaches a constant value with increasing film thickness. From eqn (2.56), we obtain,

$$\tanh^{-1}(\varepsilon q_{av}\ell/C_s \sqrt{Dk}) = \sqrt{k/D}\cdot\ell \qquad\qquad (2.59)$$

The value $C_s \sqrt{Dk}/\varepsilon$ can be estimated from eqn (2.58) by using an asymptotic value at large ℓ in the plot of $q_{av}\ell$ versus ℓ. Therefore, we can obtain $\sqrt{k/D}$ from the slope of the plot of $\tanh^{-1}(\varepsilon q_{av}\ell/C_s \sqrt{Dk})$ against ℓ. This value may be regarded as a measure of the relative contribution of the diffusion to the overall rate of oxidation. In Fig. 2.17, the rate of scission q_{av} is plotted against the film thickness 2ℓ, as expected from eqn (2.57), q_{av} approaches a constant value in the limit of thin film. The value extrapolated to $\ell = 0$ is independent of the value of the diffusion coefficient. The relation between q_{av} and ℓ is shown in Fig. 2.18. $q_{av}\ell$ is virtually constant

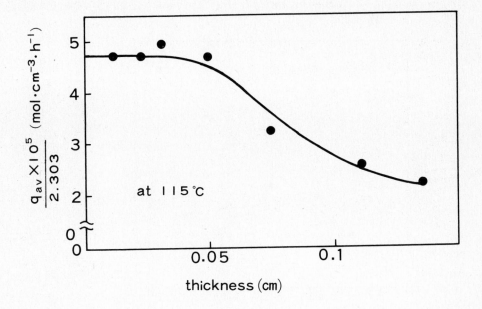

Fig. 2.17. Relation between the average rate of scission q_{av} and thickness 2ℓ.

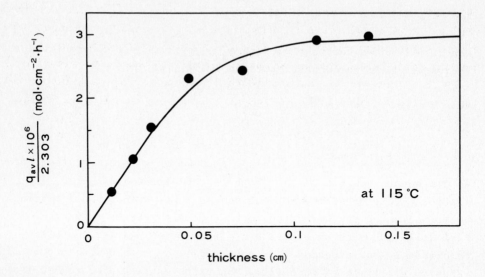

Fig. 2.18. Relation between the product $q_{av}\ell$ and thickess 2ℓ.

in the region of large ℓ, hence, the limiting value $C_S\sqrt{Dk}/\varepsilon$ may be

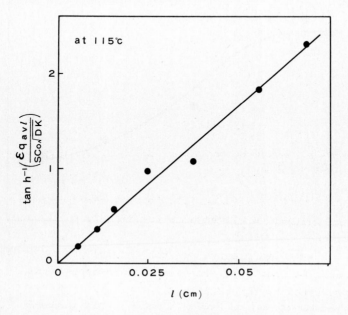

Fig. 2.19. Plot of $\tanh^{-1}(\varepsilon q_{av}\ell/C_S\sqrt{Dk})$ against ℓ.

easily determined. The plot of $\tanh^{-1}(\varepsilon q_{av}\ell/C_s\sqrt{Dk})$ versus ℓ is shown in Fig. 2.19. The experimental points are well approximated by a straight line, from the slope of which the value $\sqrt{k/D}$ is determined. The solid line in Fig. 2.18 presents the $q_{av}\ell$ versus ℓ relation cal-culated from eqn (2.56) using the $\sqrt{k/D}$ value estimated from the method described above. As shown in Figs. 2.18 and 2.19, the rate of oxidative chain scission in the vulcanized polyisoprene films is well represented by eqn (2.56) or (2.59) derived on the basis of diffusion coupled with the first-order oxygen consumption. The distribution of the concentration of O_2 in the film can also be calculated with the use of eqn (2.50). An example is shown in Fig. 2.20. The concentration

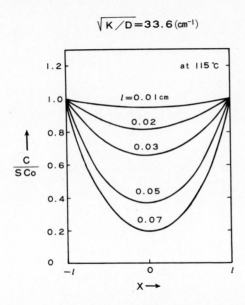

$\sqrt{K/D}=33.6\,(cm^{-1})$

Fig. 2.20. Distribution of O_2 concentration in the film for various thickness at 115°C.

of O_2 in the central part of a thick film is unexpectedly small, that is, the oxidative degradation proceeds at the expense of the surface layer in these cases.

The logarithm of $\sqrt{k/D}$ is plotted against the reciprocal of the absolute temperature in Fig. 2.21. The temperature-dependence of $\sqrt{k/D}$ may be represented by eqn (2.60).

$$\sqrt{k/D} = A\exp\left[-(E_k - E_D)/2RT\right] \qquad (2.60)$$

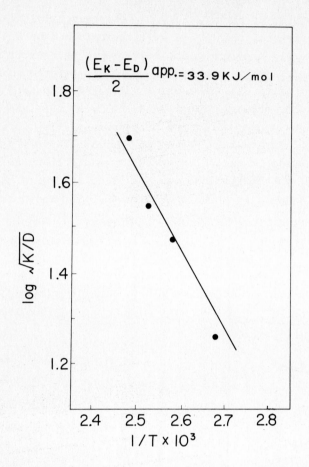

Fig. 2.21. Plot of logarithm of $\sqrt{k/D}$ vs. reciprocal absolute temperature.

where A is a constant, R is the gas constant, T is the absolute temperature, and E_k and E_D are the activation energies of the oxidation reaction and of the diffusion process, respectively. The difference $(E_k - E_D)$ is determined from Fig. 2.21 as 68 kJ mol^{-1}. Since the activation energy of the diffusion of O_2 in natural rubber vulcanizates was determined to be about 33 kJ mol^{-1}, by the time-lag method at low temperatures (ref. 26), the activation energy of the oxidation reaction is calculated to be 101 kJ mol^{-1}. This may be the true activation energy in the oxidative degradation of sulfur-cured cis-1,4-polyisoprene.

Similar treatment was carried out successfully by Billingham et al (ref. 27) with respect to oxygen absorption in crystalline poly-4-methylpentene-1. The experimental relation between O_2 uptake and the

film thickness was well approximated by an equation similar to eqn
(2.56).

As described above, quantitative estimation of the oxygen diffusion
effects on chemical stress-relaxation may be possible using rubber
samples with different film thickness. It must be emphasized that the
oxidative degradation of vulcanized rubber films is significantly
controlled by diffusion, especially at high temperatures, and that
the true activation energy can only be estimated by taking into
account the diffusion effects.

Although the steady-state experiments have enabled us to estimate
the diffusion effects, they are not capable of determining the
absolute values of D and k. This can be achieved by making use of
transient experiments (ref. 23). The outline of this method will be
shown below.

In order to obtain the expression for the O_2 concentration in the
film in the transient state, eqn (2.48) must be solved exactly. The
initial and the boundary conditions are as follows,

I. C. $t=0$, $C=0$

B. C. $\begin{cases} x=\ell, & C=C_s \\ x=0, & \partial C/\partial x=0 \end{cases}$ (2.61)

It was shown by Danckwerts (ref. 28), that the solution for such
a problem satisfies the following equation,

$$C = k \int_0^t C_1 e^{-kt} dt + C_1 e^{-kt} \qquad (2.62)$$

where C_1 is the solution to eqn (2.48) in the absence of chemical
reaction, i.e., $k=0$, for the same initial and boundary conditions.

$$\frac{\partial C_1}{\partial t} = D \frac{\partial^2 C_1}{\partial x^2} \qquad (2.63)$$

The solution to eqn (2.63) that satisfies the initial and boundary
conditions eqn (2.61) is represented by well-known trigonometric
series (ref. 24),

$$\frac{C_1}{C_s} = 1 - \frac{4}{\pi} \sum_{n=0}^{\infty} \frac{(-1)^n}{(2n+1)} e^{\frac{-D(2n+1)^2 \pi^2 t}{4\ell^2}} \cos \frac{(2n+1)}{2} \pi x \tag{2.64}$$

Substituting eqn (2.64) into (2.62) gives the solution,

$$\frac{C}{C_s} = \frac{\cosh(\sqrt{\frac{k}{D}}\cdot x)}{\cosh(\sqrt{\frac{k}{D}}\cdot \ell)} - \left[\frac{2}{\ell} \sum_{n=0}^{\infty} \frac{(-1)^n \frac{2n+1}{2} \pi}{(\frac{2n+1}{2\ell}\pi)^2 + \frac{k}{D}} e^{-\{(\frac{2n+1}{2\ell}\pi)^2 D + k\} t} \cos \frac{2n+1}{2}\pi x \right]$$

$$= \frac{\cosh(\sqrt{\frac{k}{D}}\cdot x)}{\cosh(\sqrt{\frac{k}{D}}\cdot \ell)} - \frac{2}{\ell} \sum_{n=0}^{\infty} \frac{(-1)^n D \sqrt{\alpha_{1n}}}{\alpha_{1n} D + k} e^{-(\alpha_{1n} D + k) t} \cos \sqrt{\alpha_{1n}} x \tag{2.65}$$

where,

$$\alpha_{1n} = \left(\frac{2n+1}{2\ell}\pi\right)^2 \tag{2.66}$$

When $t = \infty$, the second term of eqn (2.65) vanishes, leaving the time-independent first term which denotes the steady-state concentration within the film (eqn 2.50).

On applying Fick's first law of diffusion, the O_2 uptake per unit area of film surface $F(t)$ is given by,

$$F(t) = 2D(\frac{\partial C}{\partial x})_{x=\ell}$$

$$= 2C_s \sqrt{Dk} \tanh(\sqrt{\frac{k}{D}}\cdot \ell) + \frac{4C_s}{\ell} \sum_{n=0}^{\infty} \frac{D\alpha_{1n}}{\alpha_{1n} + \frac{k}{D}} e^{-(\alpha_{1n} D + k) t} \tag{2.67}$$

Successive integration of eqn (2.65) with respect to x and t gives the total amount of chain scission $Q(t)$ per unit volume at time t.

$$Q(t) = \frac{1}{2\ell} \int_0^t \int_{-\ell}^{\ell} q C dx dt$$

$$= \frac{k}{\varepsilon \ell} \int_0^t \int_0^{\ell} C dx dt$$

$$= \frac{C_s k}{\epsilon \ell} \left[t \sqrt{\frac{D}{k}} \tanh(\sqrt{\frac{k}{D}} \cdot \ell) - \frac{2}{\ell} \sum_{n=0}^{\infty} \frac{1-e^{-(\alpha_{1n}D+k)t}}{(\alpha_{1n}+\frac{k}{D})(\alpha_{1n}D+k)} \right] \qquad (2.68)$$

Introducing a new parameter α_n, such that,

$$\alpha_n = \alpha_{1n}D+k = (\frac{2n+1}{2\ell}\pi)^2 D+k \qquad (2.69)$$

Eqn (2.68) is reduced to,

$$Q(t) = \left[\frac{C_s \sqrt{DK}}{\epsilon \ell} \tanh(\sqrt{\frac{k}{D}} \cdot \ell) \right] t - \frac{2C_s Dk}{\epsilon \ell^2} \sum_{n=0}^{\infty} \frac{1}{\alpha_n^2}$$

$$+ \frac{2C_s Dk}{\epsilon \ell^2} \sum_{n=0}^{\infty} \frac{e^{-\alpha_n t}}{\alpha_n^2} \qquad (2.70)$$

The coefficient of the first term of the right-hand side of this equation corresponds to the average rate of chain scission q_{av} in the steady state. If we make further simplification by putting,

$$\left. \begin{array}{l} A = \dfrac{2C_s Dk}{\epsilon \ell^2} \displaystyle\sum_{n=0}^{\infty} \dfrac{1}{\alpha_n^2} \\[3mm] B(t) = \dfrac{2C_s Dk}{\epsilon \ell^2} \displaystyle\sum_{n=0}^{\infty} \dfrac{e^{-\alpha_n t}}{\alpha_n^2} \end{array} \right\} \qquad (2.71)$$

$$Q(t) = q_{av}t-A+B(t) \qquad (2.72)$$

The time-independent term A is calculated, as follows, by summing the infinite series

$$A = \frac{C_s}{\epsilon X} \left\{ \tanh(\frac{X}{2}) - \frac{X}{2} \operatorname{sech}^2(\frac{X}{2}) \right\} \qquad (2.73)$$

where X is a dimensionless parameter defined by

$$X = 2\ell \sqrt{\frac{k}{D}} \qquad (2.74)$$

The parameter X represents the degree of diffusion-control in the reaction. A high value of X means that the oxidative chain scission is highly diffusion-controlled, while a low value indicates that diffusion-control is unimportant in the reaction.

The analytical procedure for the method of transiency is schemati-
cally shown in Fig. 2.22. As described previously, the total amount
of chain-scission $Q(t)$ at time t is represented by eqn (2.72),

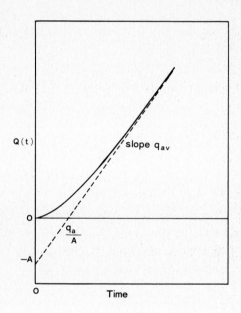

Fig. 2.22.Analytical method for the transient experiments. Eqns (2.72) and (2.75)
are represented by a solid curve and a dotted line. Intercepts on Q(t) and t axes
are −A and q_{av}/A, respectively.

$$Q(t) = q_{av}t - A + B(t) \qquad (2.72)$$

This equation is shown in Fig. 2.22 as a solid curve. As can be seen
from eqn (2.71), the curve approaches the following line at large t.

$$Q(t) = q_{av}t - A \qquad (2.75)$$

The asymptote represented by eqn (2.75) is shown in Fig. 2.22 as a
dotted line. The line has an intercept −A on $Q(t)$ axis, and the
steady-state rate of chain-scission q_{av} can be obtained from the
slope of this line. The transient response function B(t) is represent
by the difference between the solid curve and the dotted line. Althou
this function is the sum of the terms of an infinite series, the

coefficient of leading terms can readily be estimated if the series converges rapidly. The simplest case, in which $B(t)$ is expressed by a single exponential term, will be discussed below.

Using eqns (2.70) and (2.74), q_{av} is given by,

$$q_{av} = \frac{2C_s k}{\varepsilon X} \tanh(\frac{X}{2}) \qquad (2.76)$$

Dividing this equation by eqn (2.73),

$$\frac{q_{av}}{A} = \frac{1}{2k} (1-X\text{cosech } X) \qquad (2.77)$$

As shown in Fig. 2.21, this value is the intercept of the asymptote (2.75) on the t-axis, and it corresponds to "time lag" of the permeability experiments.

If $B(t)$ is approximated by a single exponential term corresponding to n=0, the coefficient α_o can be determined by the slope of the plot of $\ln[Q(t)-q_{av}t+A]$ against time, where,

$$\alpha_o = \frac{\pi^2}{4\ell^2} D+k = k(\frac{\pi^2}{X^2} + 1) \qquad (2.78)$$

Eqns (2.77) and (2.78) give a relation,

$$\text{cosech } X = \frac{\pi^2+X^2(1-2\alpha_o A/q_{av})}{X(\pi^2 + X^2)} \qquad (2.79)$$

This equation cannot be solved for X analytically, but a numerical solution is easily obtained by using a computer. Then, the absolute values of D and k are determined by applying eqn (2.77) or eqn (2.78). The relationship between $\alpha_o A/q_{av}$ and X is shown in Fig. 2.23. The solution of eqn (2.79) is alternatively obtained from this figure graphically.

The transient experiments were also carried out for sulfur vulcanizates of commercial synthetic *cis*-1,4-polyisoprene. The outline of the analysis of a stress-relaxation curve is shown in Fig. 2.24. First, the stress-relaxation is carried out in a nitrogen atmosphere. The stress-decay in N_2 is denoted by curve A and may be ascribed to a mechanism which is not associated with oxidative chain-scission, for

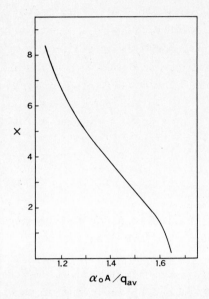

Fig. 2.23. Relationship between dimensionless numbers $\alpha_o A/q_{av}$ and X. The curve in this figure was obtained by solving eqn (2.79) numerically.

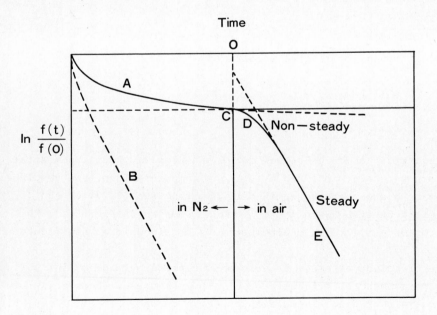

Fig. 2.24. Schematic representation for the analysis of a transient stress-relaxati curve. The atmosphere is changed instantaneously at time zero.

instance, the relaxation of trapped entanglements, exchange reactions
of sulfur linkages, etc. If the stress-relaxation is carried out in
air, the stress-decay would be faster, as denoted by a dotted line B.
At point C in Fig. 2.24, the nitrogen atmosphere is replaced
instantaneously by the air. The stress-relaxation curve passes through
a transient part D, and then gradually approaches steady-state line E.
The Q(t) vs. t relation can be calculated by using eqn (2.54). A
typical example is shown in Fig. 2.25. The network chain-density at

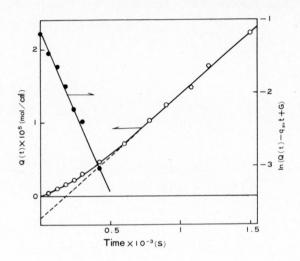

Fig. 2.25. Transient Q(t) vs. t curve for cis-1,4-polyisoprene vulcanizate at 122°C.
α_o is determined from the slope of $\ln[Q(t)-q_{av}t+A]$ vs. t plot.

the starting point of the oxidation was 9.86×10^{-6} mol cm^{-3}. The
plot of $\ln[Q(t)-q_{av}t+A]$ against time is well approximated by a
straight line, from the slope of which α_o is determined.

On applying eqn (2.79), we obtained k=1.9×10^{-3} s^{-1}, D=1.6×10^{-6}
m^{2}s^{-1}, and C_s/ε=1.5×10^{-5} mol cm^{-3}. It is interesting to compare
these values with the results obtained in the steady state experiments
(ref. 22). The value of k/D, estimated as 34 cm^{-1} from the transient
method, is in close agreement with the value 35 cm^{-1} obtained by the
steady-state method at the same temperature. This result seems to
demonstrate both the validity of applying the transient method to an
oxidation-coupled diffusion process and the validity of the original
assumptions.

The transient method, described above, can easily be extended to
the more general case (ref. 23). This is one of the applications of the

relaxation method, extensively developed by Eigen (ref. 29), and whic
may be called concentration-jump methods.

Consider an infinite film, in which a steady state is reached with
a condition of balance between the inward diffusion of oxygen and its
first-order consumption. The initial oxygen concentrations in the fil
and on the film surface for this system are denoted by Ci(x) and Cis,
respectively. If a stepwise change of O_2 concentration in the gas
phase is imposed on the system, the surface concentration instantaneo
ly attains a new value C_{fs}, while the concentration in the film C(x,t
gradually approaches a final steady-state concentration $C_f(x)$. This i
illustrated in Fig. 2.26.

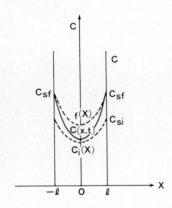

Fig. 2.26.Distribution of concentration in the film in the transient state (solid
line). $C_i(x)$ and $C_f(x)$ denote the steady-state concentrations in the initial and
final states, respectively.

Putting,

$$C(x,t) = \Delta C(x,t) + Ci(x) \qquad (2.80)$$

where $\Delta C(x,t)$ is the change (positive or negative), in the concentrat
in the film, then,

$$\frac{\partial C(x,t)}{\partial t} = D\frac{\partial^2 C(x,t)}{\partial x^2} - kC(x,t)$$

$$= D\frac{\partial^2\{\Delta C(x,t)+Ci(x)\}}{\partial x^2} - k\{\Delta C(x,t)+Ci(x)\} \qquad (2.81)$$

Since Ci(x) is the steady state concentration,

$$\frac{\partial \, Ci(x)}{\partial t} = D \, \frac{\partial^2 Ci(x)}{\partial x^2} - kCi(x) = 0 \tag{2.82}$$

Hence, eqn (2.81) reduces to,

$$\frac{\partial \varDelta C(x,t)}{\partial t} = D \, \frac{\partial^2 \varDelta C(x,t)}{\partial x^2} - k \varDelta C(x,t) \tag{2.83}$$

The mathematical form of eqn (2.83) is quite similar to that of eqn (2.48), and can be solved under the boundary condition, $\varDelta C = C_{fs} - C_{is}$ at $x = \pm \ell$, to give,

$$\frac{\varDelta C}{C_{fs} - C_{is}} = \frac{\cosh(\sqrt{\frac{k}{D}} \cdot x)}{\cosh(\sqrt{\frac{k}{D}} \cdot \ell)} - \frac{2}{\ell} \sum_{n=0}^{\infty} \frac{(-1)^n D \sqrt{\alpha_{1n}}}{\alpha_{1n} D + k} e^{-(\alpha_{1n}D+k)t} \cos \sqrt{\alpha_{1n}} x \tag{2.84}$$

The increase in O_2 uptake $\varDelta F(t)$ after the concentration jump is given by

$$\varDelta F(t) = 2(C_{fs} - C_{is}) \left\{ \sqrt{Dk} \tanh(\sqrt{\frac{k}{D}} \cdot \ell) \right.$$

$$\left. + \frac{2}{\ell} \sum_{n=0}^{\infty} \frac{D\alpha_{1n}}{\alpha_{1n} + \frac{k}{D}} e^{-(\alpha_{1n}D+k)t} \right\} \tag{2.85}$$

The corresponding equation for the increase of total amount of chain scission $\varDelta Q(t)$ is

$$\varDelta Q(t) = \frac{(C_{fs} - C_{is})Dk}{\varepsilon \ell} \left[\left\{ \tanh(\sqrt{\frac{k}{D}} \cdot \ell) \right\} t - \frac{2\sqrt{Dk}}{\ell} \sum_{n=0}^{\infty} \frac{1}{\alpha_n^2} \right.$$

$$\left. + \frac{2\sqrt{Dk}}{\ell} \sum_{n=0}^{\infty} \frac{e^{-\alpha_n t}}{\alpha_n^2} \right] \tag{2.86}$$

By applying this equation, the analytical procedures for the concentration jump methods become quite similar to that for the transient method with initial zero O_2 concentration.

The net increase in chain-scission $\Delta Q(t)$ after the concentration jump is given by,

$$\Delta Q(t) = \Delta q_{av} t - A' + B'(t) , \qquad (2.87)$$

where $\Delta q_{av} = q_{av.f} - q_{av.i}$

$$= \frac{2(C_{fs} - C_{is})k}{X} \tanh(\frac{X}{2}) \qquad (2.88)$$

where $q_{av.i}$ and $q_{av.f}$ are the steady-state rates of chain-scission in the initial and the final states. A' and $B'(t)$ are given by,

$$A' = \frac{C_{fs} - C_{is}}{X} \{ \tanh(\frac{X}{2}) - \frac{X}{2} \operatorname{sech}^2(\frac{X}{2}) \} \qquad (2.89)$$

$$B'(t) = \frac{2(C_{fs} - C_{is})Dk}{\varepsilon \ell^2} \sum_{n=0}^{\infty} \frac{e^{-\alpha_n t}}{\alpha_n^2} \qquad (2.90)$$

and then,

$$\frac{\Delta q_{av}}{A'} = \frac{1}{2k} (1 - X \operatorname{cosech} X) \qquad (2.91)$$

These equations are illustrated in Fig. 2.27.

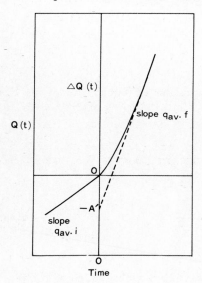

Fig. 2.27. Analytical method for the concentration jump experiments. $q_{av.i}$ and $q_{av.f}$ denote the steady-state rates of chain-scission in the initial and final states, respectively. Intercept on $Q(t)$ axis is $-A'$.

On applying eqn (2.78), we have,

$$\text{cosech } X = \frac{\pi^2 + X^2(1 - 2\alpha_0 A'/\Delta q_{av})}{X(\pi^2 + X^2)} \qquad (2.92)$$

Eqn (2.92) can be solved numerically or graphically by using Fig. 2.23, as described previously.

The principal advantage of the transient method over the steady-state method is the simplicity of the experiments. It is necessary to prepare a number of rubber films with a same network chain-density and with different thicknesses in order to estimate the parameter $\sqrt{k/D}$ by the steady-state method. By contrast, only a single film with an appropriate thickness is required in the transient method.

Diffusion coefficients of simple gases are represented by a well-known relationship (ref. 30).

$$D = D_0 \exp(-E_d/RT) \qquad (2.93)$$

where D_0 and E_d are the pre-exponential factor and activation energy for diffusion, respectively. Using this relation, D values at high temperatures can be estimated from that at a low temperature where oxidation is absent. On substituting Amerongen's data (ref. 31) of $D_0 = 1.94$ $cm^2 s^{-1}$ and $E_d = 35$ kJ mol^{-1} for natural rubber, we obtain $D = 4.9 \times 10^{-5}$ $cm^2 s^{-1}$ at 122°C. This value is considerably higher than the one obtained by the transient method.

Two reasons for this discrepancy may be advanced. First, it must be emphasized that in the case of rubbery polymers eqn (2.93) is valid only in a narrow temperature range. As shown by Amerongen (ref. 31), E_d for simple gases in natural rubber decreases with increasing temperature; this can lead to overestimation of the extrapolated D values. The second reason, and probably the dominant one, is the occurrence of a reversible reaction in the oxidation, that is,

$$R\cdot + O_2 \rightleftarrows RO_2\cdot \qquad (2.94)$$

It was suggested by Benson (ref. 32) that the ceiling temperature for this equilibrium is relatively low for R=allyl at low O_2 partial pressures. As is well known, allyl type radicals plays an important role in the oxidation of natural rubber. The presence of the reversible reaction (2.94) results in the underestimation of D values obtained by the transient method.

The above treatments of the steady-state and transient methods ar both based on the assumption of first-order consumption of O_2. More strictly, the reaction should be envisaged as bimolecular, because it is dependent upon the concentration of available sites on the molecular chains of vulcanizates. Therefore, the diffusion eqn (2.48 is to be rewritten as,

$$\frac{\partial c}{\partial t} = D \frac{\partial^2 c}{\partial x^2} - kn(x,t)c \qquad (2.95)$$

where $n(x,t)$ is the concentration of available sites for oxidative chain-scission. Unfortunately, eqn (2.95) is non-linear, and the solution can be obtained only approximately or numerically. Approxim solutions for this type of equation were obtained by Katz et al. (ref. 33), and also Reese and Eyring (ref. 34). A more complicated case, in which gas-immobilization is a reversible process, has also been discussed (refs. 35 37). The application of such treatments, however, is not necessary in the analysis of stress-relaxation curve since only one cut per network chain is sufficient to reduce the str value to zero, that is, the number of available sites for scission remains practically constant throughout the measurement.

Another problem is heterogeneity in rubbers. The presence of crystalline phases, domain structures, fillers, antioxidants, etc., complicates the situation. Recently, a theoretical treatment for the partially crystalline polymers was developed by Jellinek (ref. 38). rapid and accurate determination of the stress values as well as morphological information will be indispensable for the estimation o diffusion effects on heterogeneous rubber systems.

5 Cleavage along Main Chains
    (1) Determination of Crosslinked Network Chain-Density
    The three densities of network chains of rubbers, $n_M(0)$, $n_{SM}(0)$, and $n_S(0)$, were estimated. $n_M(0)$ was determined at 30°C in nitrogen stress-strain measurement interpreted on the basis of the statistica theory of rubber-like elasticity (ref. 39). $n_{SM}(0)$ was determined by stress-strain measurements on the swollen rubber (ref. 40). After being swollen in xylene for 72 h at 30°C, this was estimated as $\alpha=$ 1.08 1.20. Lastly, $n_S(0)$ was estimated from swelling measurements

using the Flory-Huggins equation with the Kraus correction (ref. 41). Selected pieces of the vulcanizate were accurately weighed and swollen in benzene at 30°C.

(2) Effect of Chain-Entanglements

In Table 2.3, $n_M(0)$, $N_S(0)$, and $n_{SM}(0)$ for various cross-linked natural rubbers are tabulated. It was found that values of $n_M(0)$ were usually larger than those of $n_S(0)$ or $n_{SM}(0)$.

TABLE 2.3

Preparation and characterization of crosslinked natural rubber

Sample number	DCP	Curing time,[a] min	$n_M(0)$, mole/ml × 10^{-4}	$n_S(0)$ mole/ml × 10^{-4}	$n_{SM}(0)$, mole/ml × 10^{-4}
1	2.0	10	0.66	0.25	0.14
2	2.0	20	1.10	0.44	0.23
3	2.0	30	1.25	0.68	0.36
4	2.0	40	1.48	0.83	0.52
5	3.0	30	2.22	0.97	0.68
6	3.0	65	2.35	0.78	1.02
7	3.0	80	2.46	1.60	1.10
8	5.0	20	1.96	1.28	0.84
9	5.0	30	2.72	1.85	1.55
10	5.0	40	2.89	2.40	1.56
11	5.0	45	3.15	2.20	1.88
12	10.0	30	5.13	3.01	4.16
13	10.0	40	5.53	5.06	4.23
14	10.0	50		5.51	
15	10.0	90	7.83		5.71
16	10.0	120	8.50		
17	10.0	210	8.34		8.45

[a]Curing temperature, 150°C.

The statistical theory of rubber-like elasticity states that the tensile stress needed to stretch a rubber is proportional to the number of network chains per unit volume. It has been further shown that a correction was necessary to account for loose chain-ends in the rubber (ref. 42).

In this chapter, another correction, the effect of chain-entanglements, is discussed. The entanglements which are induced during crosslinking will act, in many respects, as if they were true chemical crosslinks. $n_S(0)$ and $n_{SM}(0)$ will indicate accurately the numbers of chemically crosslinked chains but $n_M(0)$ will also include a contribution to chain-entanglement from physically interacting chains. Therefore, the result in Table 2.3 is consistent with to prevailing view. Two important relations—$n_S(0)$ versus $n_M(0)$ and $n_{SM}(0)$ versus $n_M(0)$—are

48

shown in Fig. 2.28. This figure indicates that each should be a

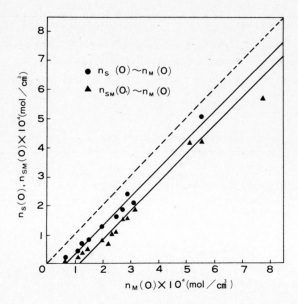

Fig. 2.28. Relations of $n_S(0)$ vs. $n_M(0)$ and $n_{SM}(0)$ vs. $n_M(0)$.

straight line of unit slope for the overall range of n(0) except for very low degrees of crosslinking. $n_S(0)$ and $n_{SM}(0)$ measure the same parameter, but the relation of $n_S(0)$ versus $n_M(0)$ is not consistent with that of $n_{SM}(0)$ versus $n_M(0)$. This presumably can be attributed to the differences in the swelling solvents and experimental methods employed. This suggests that the number of effective chain entanglements is constant and independent of network chain-density n(0) at relatively high degrees of crosslinking.

Therefore,

$$n_M(0) = n_I(0) + n_{II}(0) = n_I(0) + \text{constant} \tag{2.96}$$

where $n_I(0)$ is $n_S(0)$ or $n_{SM}(0)$ representing the chemical network chain-density and $n_{II}(0)$ is the number of effective chain-entanglements. This result is consistent with the work of Mullins (ref. 43).

Mullins' data were obtained on natural rubber at low degrees of crosslinking. Our result indicates that eqn (2.96) also applies to

high degrees of crosslinking. Since the intercept of the straight
lines of unit slope is $n_{II}(0)$, according to eqn (2.96), it is evident
from Fig. 2.28 that $n_{II}(0) \approx (0.7\sim1.0) \times 10^{-4}$ mol cm^{-3}. This leads to the
the supposition that the molecular weight between entanglement points
(Me) will be about $9,000\sim12,000$ which is in agreement with Mullins'data
and does not differ too greatly from the value found for uncrosslinked
natural rubber. It becomes doubtful, however, whether such a relation
as eqn (2.96) between $n_M(0)$ and $n_S(0)$ or $n_{SM}(0)$ will apply at higher
$n(0)$. Actually, from recent studies on SBR (ref. 44), chloroprene
rubber (ref. 45), and natural rubber vulcanizates, it was found tht
the region of $n_M(0) \langle n_S(0)$ occurs at very high degrees of crosslinking.
This indicated by Fig. 2.29 and 2.30.

From these figures it will be observed that these curves show a

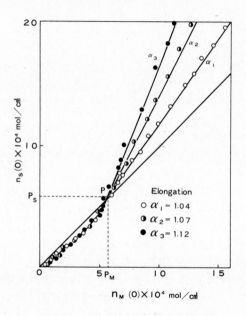

Fig. 2.29.Relation between $n_M(0)$ and $n_S(0)$ for SBR vulcanizates.

tendency for $n_M(0) \rangle n_S(0)$ below a certain point, P (for example,
$n_M(0) \approx 5.5 \times 10^{-4}$ mol cm^{-3} for SBR vulcanizates, and $n_M(0) \approx 4.2 \times 10^{-4}$
mol cm^{-3} for natural rubber vulcanizates); once over this point, they
indicate a tendency for $n_M(0) \langle n_S(0)$, independent of the kind of
polymers and solvents used for the swelling measurement. The reason
why $n_S(0) \rangle n_M(0)$ in the range of $n(0) \rangle P$ has recently been made clear
by our calculations (ref. 46) derived from the theory of Smith (ref.

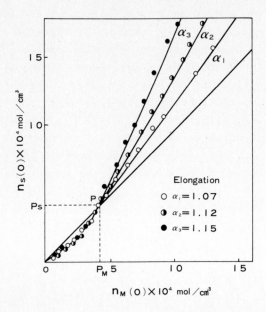

Fig. 2.30. Relation between $n_M(0)$ and $n_S(0)$ for natural rubber vulcanizates.

47) for tightly crosslinked polymers.

(3) Chemorheology of Natural Rubber Vulcanizates Cured by Various Curing Methods (ref. 48)

Natural rubber (NR-RSS-1) was cold-milled with curing ingredients under conditions described in Table 2.4. Sheets (about 0.5 mm) of mil polymer were pressed at 200 kg cm^{-2} and 145°C.

Samples 1 and 3, respectively, were dicumyl peroxide (DCP)-cured NR, and tetramethyl thiuram disulfide (TMTD)-cured NR. It is evident that the crosslinking site consists of carbon-carbon bonds in both samples, mono- and di-sulfide linkages together with for samples 1 and 3, respectively. All other samples were prepared by γ-radiation from a ^{60}Co source at room temperature, so the crosslink in sample 2 is due to carbon-carbon bonding. Sample 4 was prepared by exposing sample 3 to γ-rays. Sample 5 was prepared as follows: thin sheets (about 0.5 mm) of non-crosslinked rubber containing TMTD were made by milling and hot-pressing, and the sheet was exposed to γ-rays. It can be assumed that the crosslinks in samples 4 and 5 are mixtures of carbon-carbon bonds and mono- and di-sulfide linkages. All sample were extracted with hot acetone for 48 h and dried in vacuo.

As described above, the crosslinkages in both samples 1 and 2 are

TABLE 2.4

Preparation of cured NR polymers

	Sample				
	1	2	3	4	5
Rubber	100	100	100	100	100
Sulphur	–	–	–	–	–
Zinc oxide	–	–	5	5	–
Stearic acid	–	–	2	2	–
TMTD	–	–	3	3	3
Mercapto benzo-thiozole (MBT)	–	–	–	–	–
DCP	3	–	–	–	–
Hot-press curing*	10 min.	–	10 min.	10 min.	–
Irradiation curing+	–	+	–	+	+

* Curing temperature 145°C.
+ Total dose=12, 28.8 and 43.2 Mrad.

known to be carbon-carbon bonds. If there should be no difference of chemical and physical structure between samples 1 and 2, the chemical and physical stress-relaxation for both samples under the same conditions should be represented by a single curve. Stress-relaxation for samples 1 and 2 (having the same density $N(0) \approx 0.65 \times 10^{-4}$ mol cm^{-3}) was measured in both air and nitrogen at 100°C. The relation between the relative stress $f(t)/f(0)$ and the logarithm of time (log t) is shown in Fig. 2.31. It can be seen from this figure that the

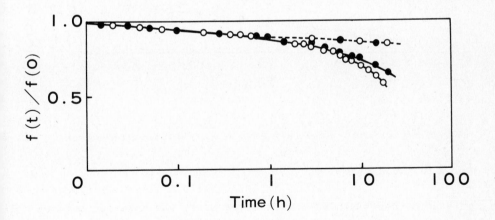

Fig. 2.31.Stress-relaxation of samples 1 (o) and 2 (●) in air and under nitrogen at 100°C, ⎯⎯ in air; ---- under nitrogen.

relaxation for sample 1, both in air and under nitrogen, is very closely consistent with that for sample 2. This suggests that sample 1 and 2 have identical chemical and physical structures in spite of the difference in curing methods. However, the stress-decay in air of sample 1 is slightly faster than that of sample 2 of long times, as can be seen in Fig. 2.31. This assures because the scission of main chains will be more facile in sample 1 because of peroxide impurities. As we shall now discuss the region where the two curves agree, it can be assumed that there is little difference between samples 1 and 2.

The stress-relaxation of three samples of type 1 having different initial chain densities $N(0)$, and prepared by the alteration of irradiation dose, were measured in air at 100°C. These stress-relaxation curves, given in Fig. 2.32, show that the rate of stress-

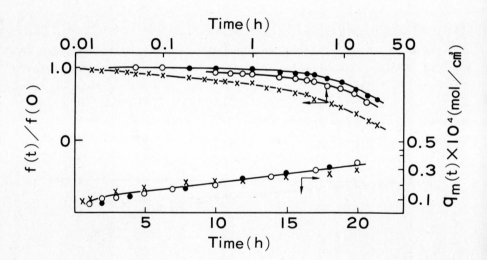

Fig. 2.32. Stress-relaxation of sample 1 specimens having different initial chain-density, and relation between the number of moles of main-chain-scission $q_m(t)$ and time t.
x, Irradiation dose = 120 kJ kg^{-1}, $N(0)=0.6 \times 10^{-4}$ mol cm^{-3}; o, 288 kJ kg^{-1}, 1.25×10^{-4} mol cm^{-3}; ●, 432 kJ kg^{-1}, 1.53×10^{-4} mol cm^{-3}.

relaxation decreases with increasing $N(0)$. The relation between $q_m(t)$ and t is also shown in Fig. 2.32. The fact that $q_m(t)$ is independent of $N(0)$ is typical oxygen-induced cleavage of the main chain, as is shown in section 2.4 below.

Fig. 2.33 displays the chemical stress-relaxation curves (log

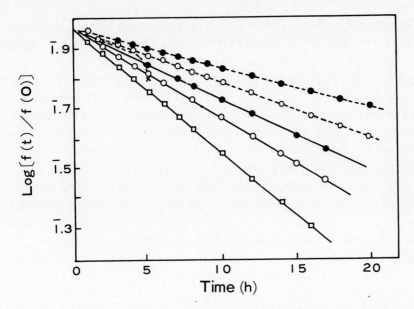

Fig. 2.33. Stress-relaxation of samples 3, 4 and 5 in air at 100°C. x, Sample 3: N(0)=1.31x10^{-4} mol cm^{-3}. ----, Sample 4: ○, irradiation dose = 288 kJ kg^{-1}, N(0)= 1.63x10^{-4} mol cm^{-3}; ●, 432 kJ kg^{-1}, 2.10x10^{-4} mol vm^{-3}. ——, Sample 5: □, 120 kJ kg^{-1}, 0.79x10^{-4} mol cm^{-3}; ○, 288 kJ kg^{-1}, 1.03x10^{-4} mol cm^{-3}; ●, 432 kJ kg^{-1}, 1.21x10^{-4} mol cm^{-3}.

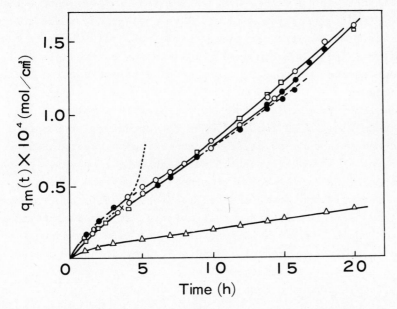

Fig. 2.34. Relation between $q_m(t)$ and t, Samples 2, 3, 4 and 5. △, Sample 2; x, Sample 3. ----, Sample 4: ○, irradiation dose = 288 kJ kg^{-1}, N(0)=1.63 10^{-4} mol cm^{-3}; ●, 432 kJ kg^{-1}, N(0)=2.10x10^{-4} mol cm^{-3}. ——, Sample 5: □,120 kJ kg^{-1}, N(0)=0.79x10^{-4} mol cm^{-3}; ○, 288 kJ kg^{-1}, N(0)=1.03x10^{-4} mol cm^{-3}; ●, 432 kJ kg^{-1}, N(0)=1.21x10^{-4} mol cm^{-3}.

f(t)/f(0) versus time) for samples 4 and 5 in air at 100°C. The stre
decay of these samples is also dependent on N(0), as in the case of
sample 1. Further linear relationship is seen for all samples over
the longer time region; it is assumed, therefore, that only the main
chain undergoes oxidative scissions in these curing systems.

The numbers of moles of main chain scission in these samples, q_m(
were calculated from eqn (2.27). The relation between q_m(t) versus
time is shown in Fig. 2.34. The value of q_m(t) is almost independent
of N(0) suggesting that the oxidative scission in both samples 4 and
5 also occurs on the main chain. What is also interesting in Fig. 2.
is that the results are independent of the conditions of sample
preparation.

The crosslinkages of samples 4 and 5, as described above, are
expected to consist of carbon-carbon bonds and mono- and di-sulphide
linkages. Now, the effective chain densities N_c(0) and N_m(0), respec
tively, are assumed to correspond to the carbon-carbon crosslinkages
and mono-, and di-sulphide linkages. In this case, N(0) is equal to
N_c(0)+N_m(0). As no experimental method for the estimation of the
values of N_c(0) and N_m(0) has so far been made known, we connot
evaluate them. The ratio, ρ of N_c(0) to N_m(0) may be different in
samples 4 and 5. The ρ value, even in the same sample, can be expect
to change with the irradiation dose. Nevertheless the relations of
q_m(t) versus t are consistent among themselves. This suggests that
both samples 4 and 5 undergo oxidative scission on the main chain
independent of the ρ value, i.e. of the chemical structure of the
crosslinkages.

The result of sample 3 having N(0) $\approx$ 1.31 x 10^{-4} mol cm^{-3} is also
shown in both Figs. 2.33 and 2.34. The crosslinkages of sample 3
consist of mono- and di-sulphide linkages. The stress-relaxtion
mechanism for this sample is still not elucidated, as shown by the
different opinions expressed by Tobolsky (ref. 8) and Watson et al.
(ref. 2). Recently, Murakami and Tamura (ref. 49-51), and Takahashi
and Tobolsky (ref. 52) reported that chemical stress-relaxation in
air showed that oxidative scissions along the main chains were the
major causes of stress decay for sample 3. They do not, however,
provide sufficient experimental evidence for this suggestion. From
Fig. 2.34, it is apparent that the q_m(t) of sample 3 is consistent
with that of samples 4 and 5 up to about 4 h. In the longer time
region (after 4 h), it deviates from samples 4 and 5. The rapid
stress-decay probably reflects physical relaxation based on micro-
and/or macro-cracks formed in sample 3. So, we are not able to give

the chemorheological analysis in the longer time region. From the
above, it can be concluded that the chemical stress-relaxation up
to 4 h for sample 3 is based upon oxidative scission along the main
chain. Once again, it is cocluded that only main chain scission occurs
in these curing systems, and that $q_m(t)$ is independent of $\rho = N_c(0)/N_m(0)$.

However, this does not necessarily imply that $q_m(t)$ is independent
of ρ over the wide range from zero to infinity. The $q_m(t)$ of samples
1 and 2 for $\rho = \infty$ is appreciably smaller than that of samples 3, 4 and
5, as shown in Fig. 2.34, because oxidative scission of the main chain
is probably accelerated by having mono- or di-sulphide crosslinkages
present, although the acceleration mechanism is still not understood.

(4) Chemorheological Treatment of Dicumyl Peroxide-cured Natural
Rubber

On the basis of the discussion in the preceding section, we investigated the chemorheology of crosslinked natural rubber (ref. 53). The
stress-relaxation of crosslinked natural rubber was measured at 109°C
in air. The stress-relaxation curves for the samples in Table 2.3 are
shown in Fig. 2.35.

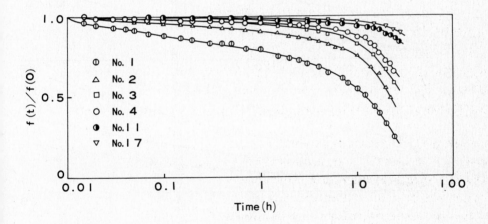

Fig. 2.35. Stress-relaxation curves of crosslinked natural rubbers at 109°C in air.

Assuming that chain-scission occurs randomly throughout the rubber
network, eqn (2.97) is derived as described before for eqn (2.21).

$$q_m(t) = -n_M(0)\ln f(t)/f(0) = -n_M(0)\ln n_M(t)/n_M(0) \qquad (2.97)$$

In this rubber, if scission occurs only along the network chains,
the relation of $q_m(t)$ versus t must be shown by a curve independent

of $n_M(0)$. Our results indicate, however, that such a relation can be
shown by a number of curves which do depend on $n_M(0)$. So, considerin
the effect of chain-entanglements as described in the previous secti
we propose eqns (2.98) and (2.99) as modifications of eqn (2.97).
These equations give the number of real main chain scissions, for th
chain-entanglements have been removed:

$$q_S(t) = -n_S(0)\ln n_S(t)/n_S(0) \tag{2.98}$$

$$q_{SM}(t) = -n_{SM}(0)\ln n_{SM}(t)/n_{SM}(0) \tag{2.99}$$

The relations $q_S(t)$ versus t and $q_{SM}(t)$ versus t are shown in Fig
2.36.

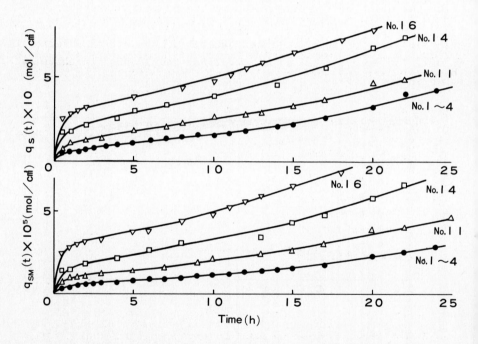

Fig. 2.36. $q_S(t)$ and $q_{SM}(t)$ vs. t curves estimated by eqns (2.98), (2.99) and (2.1

For small $n(0)$ (samples 1 to 4), this relation is represented by
one curve; that is, scission of dicumyl peroxide-cured natural rubbe
occurs at the main chain. At large values of $n(0)$, this relation wil
deviate from the curve, and it is shown that the greater $n_S(0)$ or
$n_{SM}(0)$, the greater the $q_m(t)$. The cause of this deviation seems to
be closely connected with the relation between $n_M(0)$ and $n_S(0)$ in th

range of $n(0) > P$.

Next, scission in uncrosslinked rubber was followed by determination of intrinsic viscosity for a rubber solution in toluene, oxidized at 109°C. The relationship between $[\eta]$ and molecular weight $\bar{M}_n$ is given by

$$[\eta] = 5.00 \times 10^{-4} \bar{M}_n^{0.67} \tag{2.100}$$

The quantity $q_m(t)$ was determined by

$$q_m(t) = d[1/\bar{M}_n(t) - 1/\bar{M}_n(0)] \tag{2.101}$$

where d is the natural rubber density, $\bar{M}_n(0)$ is the initial number-average molecular weight, and $\bar{M}_n(t)$ is the number-average molecular weight at later times. Lastly, the various quantities q(t) estimated by eqns (2.98), (2.99) and (2.101) are shown in Fig. 2.37. These curves are approximately consistent for samples 1 to 4.

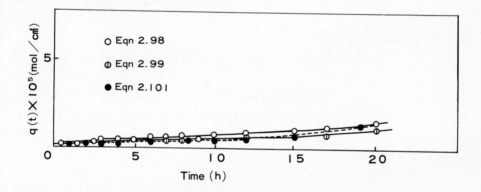

Fig. 2.37. $q_m(t)$ vs. t curves estimated by eqns (2.98), (2.68) and (2.101).

From the above results, it is obvious that the experimental values of $q_m(t)$, $q_S(t)$ and $q_{SM}(t)$ become unreliable for samples of higher $n(0)$.

(5) Rubber Elasticity on Chemical Relaxation

The modified equation of rubber elasticity (2.2), suggested by Tobolsky, has been used as the basis of chemorheology. We have discussed the applicability of this relationship, and whether the degraded stress f(t) calculated from this formula would be consistent with the initial stress $f_t(0)$ of another polymer if the degraded crosslinking density values $n(t)$, as shown in Fig. 2.38(b), are equal to the initial densi values $n_t(0)$ in Fig. 2.38(a) of another polymer.

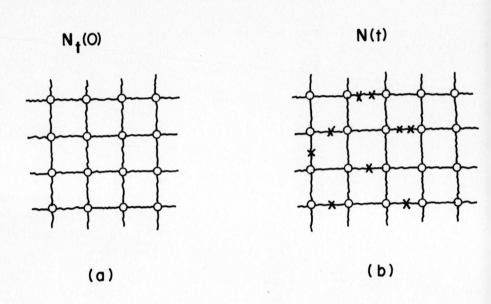

$$N_t(0) \qquad\qquad N(t)$$

$$(a) \qquad\qquad\qquad (b)$$

Fig. 2.38. (a) Structure of ideal network $n_t(0)$. (b) Degraded network $n(t)$.

It is well known that the expression for the tensile stress to deform a rubber to an extension ratio α is given by

$$f(0) = n(0)RT(\alpha - \alpha^{-2}) \tag{2.1}$$

Eqn (2.1) was modified and derived by Tobolsky as

$$f(t) = n(t)RT(\alpha - \alpha^{-2}) \tag{2.2}$$

which is regarded as one of the basic relationships in chemorheology. First, assume $n_M(t) \neq n_{M,t}(0)$, and consider the following general equation when $n(t) = n_t(0)$:

$$n_M(t) = \sigma(t) \cdot n_{M,t}(0) \tag{2.102}$$

In discussing the validity of eqn (2.2), the values of $q(t)$ which car be obtained from eqn (2.21) using eqn (2.2) are considered as apparent values:

$$q_{app}(t) = -n_{M,0}(0)\ln \{n_M(t)/n_{M,0}(0)\} \tag{2.103}$$

On the other hand, from this, the intrinsic values $q_{real}(t)$ are expressed by eqns (2.104) and (2.105):

$$q_{real}(t) = -n_{M,0}(0)\ln \{n_{M,t}(0)/n_{M,0}(0)\} \tag{2.104}$$

which is based on the following swelling method:

$$q'_{real}(t) = -n_S(0)\ln \{n_S(t)/n_S(0)\} \tag{2.105}$$

From eqns (2.102), (2.103), and (2.104), we can calculate

$$q_{app}(t) = -n_{M,0}(0)\ln \sigma(t) -n_{M,0}(0)\ln \{n_{M,t}(0)/n_{M,0}(0)\}$$

$$= -n_{M,0}(0)\ln \sigma(t) + q_{real}(t) \tag{2.106}$$

From eqn (2.106),

$$\sigma(t) = \exp \left\{ \frac{q_{real}(t)-q_{app}(t)}{n_{M,0}(0)} \right\} \tag{2.107}$$

In eqn (2.107), using $q'_{real}(t)$ of eqn (2.105) instead of $q_{real}(t)$, we obtain,

$$\sigma'(t) = \exp \left\{ \frac{q'_{real}(t)-q_{app}(t)}{n_{M,0}(0)} \right\} \tag{2.108}$$

By studying the relation between time and $\sigma(t)$, values for the oxidative degradation process in air at 100°C of the three previous SBR samples were obtained and are shown in Fig. 2.39. In Fig. 2.39, the relation between $\sigma(t)$ and t shows the straight line of $\sigma(t)=1$ to be independent of the elapsed time through the range of samples. The fact that the relation $n_M(t)=n_{M,t}(0)$ is substantiated by another paper (ref. 54) makes it a reasonable result. However, the relation between $\sigma'(t)$ and t, as seen in the figure, is a descending curve along the elapsed t

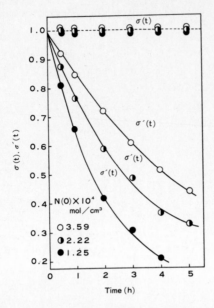

Fig. 2.39. Relation between $\sigma(t)$, $\sigma'(t)$ and time for SBR vulcanizates.

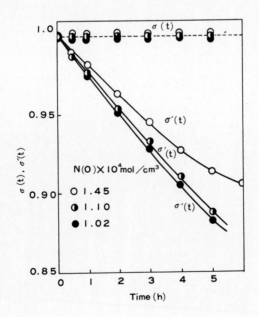

Fig. 2.40. Relation between $\sigma(t)$, $\sigma'(t)$ and time for natural rubber vulcanizates.

axis. This results from values of $n_S(t)$ and not of $n_M(t)$ for $q'_{real}(t)$ because, as previously mentioned, $n_M(t) > n_S(t)$.

In the case of cured natural rubbers, the relation between $\sigma(t)$ and t can be plotted as shown in Fig. 2.40. As in Fig. 2.39, $\sigma(t)=1$ is independent of t, $\sigma(t)$ follows a line parallel to the abscissa, and $\sigma'(t)$ follows a descending curve, for reasons that can be considered the same as in the case of cured SBR.

II SCISSION AT CROSSLINKAGES

Suppose a scission occurs only at a crosslink site in the polymer network; we represent the number of initial crosslinkages by $C(0)$, the number of scissions after time t by $q_c(t)$, and the proportionality constant by k; then the following equation can be established.

$$\frac{dq_c(t)}{dt} = k \{ C(0)-q_c(t) \} \qquad (2.109)$$

Solving eqn (2.109), we obtain,

$$q_c(t) = C(0) \cdot (1-e^{-kt}) \qquad (2.110)$$

If $t(t)/f(0)$ is the relative stress and $n(0)$ the initial chain-density if scission occurs only at the crosslinks, then

$$f(t)/f(0) = 1 - \{ 2q_c(t)/n(0) \} \qquad (2.111)$$

In an ideal network of chains,

$$2C(0) = n(0) \qquad (2.112)$$

Therefore,

$$f(t)/f(0) = e^{-kt} \qquad (2.113)$$

When scission occurs only at the crosslink in crosslinked polymers, it is obvious from eqn (2.113) that the stress-relaxation curve is expressed by a typical Maxwellian decay.

III INTERCHANGE REACTION

62

The stress-relaxation of polysulfide rubber in an atmosphere of n-butylmercaptan is more rapid than that in air, as is shown in Fig 2.41. From this fact and for other reasons, polysulfide rubber is

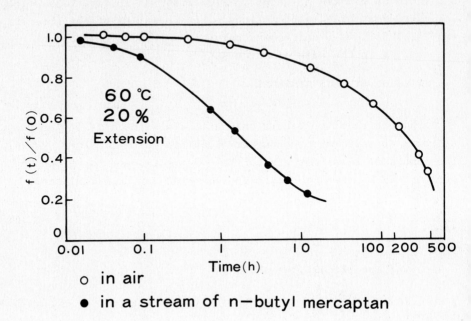

o in air

● in a stream of n−butyl mercaptan

Fig. 2.41. Chemical stress-relaxation curves for polysulfide rubbers in an atmosphere of n-butyl mercaptan.

inferred to react with n-butylmercaptan in the following way.

R-S-S-R + Bu-S-H → R-S-H + Bu-S-S-R

In more detail, polysulfide rubbers are rubber networks composed of polysulfide chains of the following structural types (refs. 55, 56):

—$\wedge\wedge\wedge$-CH$_2$CH$_2$SSCH$_2$CH$_2$SS-$\wedge\wedge\wedge$—

ethyl disulfide polymer

—$\wedge\wedge\wedge$-CH$_2$CH$_2$OCH$_2$CH$_2$SSCH$_2$CH$_2$OCH$_2$CH$_2$SS-$\wedge\wedge\wedge$—

diethylether disulfide polymer

—ⱱⱱ— $CH_2CH_2OCH_2OCH_2CH_2SSCH_2CH_2OCH_2OCH_2CH_2SS$ —ⱱⱱ—

diethylformal disulfide polymer

Although the disulfide linkages are predominant, there are also monosulfide linkages $-S-$, trisulfide linkages $-SSS-$, and tetrasulfide linkages $-SSSS-$ along the main polymer chains. The network junctures are generally trifunctional covalent linkages. The ends of terminal network chains are SH linkages.

In addition to these major features of the network structure, there are minor amounts of linkages of the type $-S^{=}---^{+}Pb^{+}---^{=}S-$ along the chains and $-S^{=}----Na^{+}$ at the ends of the terminal network chains.

Chemical interchanges occur in the network, and become manifest if the rubber is maintained at a fixed strain for an appreciable length of time. The chemical interchanges are of the following kinds:

$$+\begin{matrix} R\ -S-S-R \\ R'-S-S-R' \end{matrix} \xrightarrow{cat.} \begin{matrix} R-S \\ R'-S \end{matrix} + \begin{matrix} S-R \\ S-R' \end{matrix}$$

$$+\begin{matrix} R\ -S-H \\ R'-S-S-R' \end{matrix} \xrightarrow{cat.} \begin{matrix} R\ -S \\ R'-S \end{matrix} + \begin{matrix} H \\ S-R' \end{matrix}$$

$$+\begin{matrix} R\ -S-Na \\ R'-S-S-R' \end{matrix} \longrightarrow \begin{matrix} R\ -S \\ R'-S \end{matrix} + \begin{matrix} Na \\ S-R' \end{matrix}$$

$$+\begin{matrix} R\ -S-Pb-S-R \\ R'-S-S-R' \end{matrix} \longrightarrow \begin{matrix} R\ -S-Pb \\ R'-S \end{matrix} + \begin{matrix} S-R \\ S-R' \end{matrix}$$

where R and R' represent polymer chains.

Trisulfide and tetrasulfide linkages are also involved in these bond interchanges. The catalysts indicated in the above equations are ionic substances (acids, bases, or salts) left over from the poly-merization or vulcanization process.

For a stretched rubber sample, the elementary process of stress-decay is indicated by Fig. 2.42.

Polymers other than polysulfide rubbers also undergo interchange reactions. For example, the silicone rubbers whose basic chain structure is

$$\begin{matrix} & CH_3 & CH_3 & CH_3 & CH_3 \\ & | & | & | & | \\ \text{—ⱱⱱ—} & Si\ -O-Si\ -O-Si\ -O-Si\ -O & \text{—ⱱⱱ—} \\ & | & | & | & | \\ & CH_3 & CH_3 & CH_3 & CH_3 \end{matrix}$$

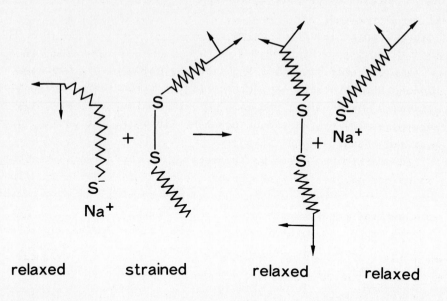

relaxed   **strained**   **relaxed**   **relaxed**

Fig. 2.42. Mechanism of interchange reaction.

also can undergo interchange reactions of the type

$$R-O-Si-R \\ + \\ R'-Si-O-R' \quad \xrightarrow{cat.} \quad R-O \quad Si-R \\ \quad | \; + \; | \\ R'-Si \quad O-R'$$

where the catalyst is an acid, base or salt left over from polymeri-
zation or vulcanization (refs. 57,58).

The interchanges considered need not be along the chain but may
occur at the crosslinks. Certain chemical crosslinks, such as salt
linkages, are very labile and may be involved in interchange reaction
The interchange in this case may, of course, be regarded as a two-sta
process, the breaking and remaking of crosslinks, where the breaking

$$\begin{matrix} & O & & & & O & \\ & \| & & & & \| & \\ -C-O^{-} & ------- & Me^{++} & ----- & ^{-}O-C- \end{matrix}$$

and re-forming of bonds occur at equal rates. Furthermore, temporary
"physical" crosslinks that break and remake may be formed between
polymer chains, and can play the same role as labile chemical crosslinks. Such a situation may possibly occur in a polymer containing a

few functional groups that could give rise to strong points of interaction with neighbouring chains, e.g., a hydrogen bond or a salt
linkage. These types of linkages may actually be encountered in
swollen wool fibers in addition to the covalent -SS- crosslinkages,
and they may well be encountered in carboxyl-containing polymers. In
general, however, the long-range physical interactions among high-
molecular-weight linear polymers are best described as complex
"entanglement". The "entanglements" act as temporary crosslinks that
break and remake.

Immediately after stretching from the original length, the stress
$f(0)$ is given by

$$f(0) = N(0)kT(\alpha - \alpha^{-2}) \tag{2.1}$$

where $N(0)$ is the number of effective network chains per cubic centimeter of rubber at time zero.

If the rubber is maintained at constant length, at time t only
$N(t)$ of the $N(0)$ network chains per cubic centimeter will not have
been affected by the interchange. We shall assume that, if a chain
has engaged in bond-interchange, it is relaxed with respect to the
external length of the sample, and will no longer contribute to the
stress in a sample maintained at that length. Only the $N(t)$ network
chains that have not undergone bond-interchange will contribute to
the stress, and therefore,

$$f(t) = N(t)kT(\alpha - \alpha^{-2}) \tag{2.2}$$

The interchange reaction might be expected to obey the following
rate law.

$$- \frac{dN(t)}{dt} = k'n_1m_2N(t) \tag{2.114}$$

where n_1 is the average number of linkages along each chain which are
available for interchange (e.g., the total number of disulfide,
trisulfide, and tetrasulfide linkages), m_2 is the number of S-Na
terminal groups per cubic centimeter of rubber, and k' is a specific
rate constant. The boundary condition is $N(t)=N(0)$ at $t=0$.

Integrating eqn (2.114), one obtains,

$$N(t) = N(0) \cdot e^{-kt} = N(0) \cdot e^{-t/\tau} \qquad (2.115)$$

where $k = n_1 m_2 k'$, $\tau = 1/k$

Substituting eqn (2.115) into eqn (2.2) and dividing the result by eqn (2.1), we find

$$\frac{f(t)}{f(0)} = e^{-kt} = e^{-t/\tau} \qquad (2.116)$$

Eqn (2.116) indicates that the relaxation caused by interchange reactions can also be expressed by Maxwellian decay.

The above results are based upon the assumption that there are $\bar{x}_m$ interchangeable linkages per chain and that they all have the sa facility for interchange, but that the probability of reaction of any individual groups is randomly determined.

We shall assume that there are N_0, N_1, N_2, -------$N_{\bar{x}_m -1}$ and $N_{\bar{x}_m}$ network chains per unit volume containing 0, 1, 2, ---- $\bar{x}_m -1$ and $\bar{x}_m$ interchangeable linkages, respectively, and that the probability of interchange is P_ω.

Let us consider the total number of chains to be $N(0)$, then the relation between N_i and $\bar{x}_i$ is given by (ref. 59),

$$N_0 = N(0) \cdot (1-P_\omega)^{\bar{x}_m}$$
$$N_1 = N(0) \cdot \bar{x}_m \cdot (1-P_\omega)^{\bar{x}_m -1} \cdot P_\omega$$

$$N_2 = N(0) \frac{\bar{x}_m \cdot (\bar{x}_m -1)}{2} \cdot (1-P_\omega)^{\bar{x}_m -2} P_\omega^2 \qquad (2.117)$$

$$N_{\bar{x}_m} = N(0) \cdot P_\omega^{\bar{x}_m}$$

The total number of network chains which remain after time t, $N(t)$, is represented by $N(t) = \sum\limits_{i=0}^{\bar{x}_m} N_i$ and P_ω is very small. In that case terms other than N_0, and N_1 in eqn (2.117) are negligible. Accordingly,

$$\frac{f(t)}{f(0)} = (1-P_\omega)^{\overline{x}_m} \cdot \exp(-k \cdot \overline{x}_m \cdot t) + \overline{x}_m (1-P_\omega)^{\overline{x}_m-1} \cdot P_\omega \cdot \exp\{-[k(\overline{x}_m-1)t+k_\omega t]\} \quad (2.118)$$

where k is the rate constant of the normal interchange reaction and k_ω is that of a weak interchange reaction.
Eqn (2.118) is simplified as follows,

$$\frac{f(t)}{f(0)} = A \cdot e^{-k'r} + B \cdot e^{-k''t} \qquad \text{where } A + B = 1 \qquad (2.119)$$

Strictly speaking, an interchange reaction was found to be expressed by the summation of two items of Maxwellian decay, as shown by eqn (2.119) rather than eqn (2.116). A theoretical study on the mechanism of interchange reactions has been carried out by Baxter et al. (refs. 60,61).

Let us consider a system that has been vulcanized at random and then degraded so that crosslink scissions and recombinations occur. If γ is the probability that any specific monomer unit will be joined to a crosslink which exists at time t, then the probability of finding a chain of length x units, i.e. one in which (x-1) units are left unvulcanized at time t will be $\gamma(1-\gamma)^{x-1}$. The number of chains of length x units N_x at time t will therefore be

$$N_x = N(t) \cdot \gamma(1-\gamma)^{x-1} \qquad (2.120)$$

where N(t) is the total number of chains. If there are C(t) crosslinks in the system, then

$$\gamma = \gamma(0) \frac{C(t)}{C(0)} \qquad (2.121)$$

where $\gamma = \gamma(0)$ and C(t)=C(0) when t=0.
Ignoring the effects of chain ends,

$$\frac{N(t)}{N(0)} = \frac{C(t)}{C(0)} \qquad (2.122)$$

where N(0) is the total number of chains at time 0. Therefore

$$N_x = N(0) \cdot \gamma(0) \cdot \{\frac{C(t)}{C(0)}\}^2 \cdot \{1-\gamma(0) \cdot \frac{C(t)}{C(0)}\}^{x-1} \tag{2.123}$$

Simultaneously with the crosslinking reaction, scission of hydrocarbon chains is assumed to take place at random. If p is the probability of scission up to time t then the probability of survival of chain of length x will be $\gamma(1-\gamma)^{x-1} \cdot (1-p)^{x-1}$, since $(1-p)$ will be the probability of scission in a monomer residue not occurring. The number of chains of length x existing at time t will therefore be

$$N_x = N(0) \cdot \gamma(0) \cdot \{\frac{C(t)}{C(0)}\}^2 \cdot \{1-\gamma(0) \cdot \frac{C(t)}{C(0)}\}^{x-1} (1-p)^{x-1} \tag{2.124}$$

The total number of chains will then be

$$N(t) = \sum_{x=1}^{\infty} N(0) \cdot \gamma(0) \cdot \{\frac{C(t)}{C(0)}\}^2 \cdot \{1-\gamma(0) \cdot \frac{C(t)}{C(0)}\}^{x-1} \cdot (1-p)^{x-1}$$

$$= \frac{N(0) \cdot \gamma(0) \cdot \{\frac{C(t)}{C(0)}\}^2}{\gamma(0) \cdot \frac{C(t)}{C(0)} + p} \tag{2.125}$$

Assuming that the kinetic theory of elasticity applies in this system, then

$$\frac{f(t)}{f(0)} = \frac{N(t)}{N(0)} \tag{2.3}$$

Accordingly, for a given elongation, the relative stress will be

$$\frac{f(t)}{f(0)} = \frac{\gamma(0) \cdot \{\frac{C(t)}{C(0)}\}^2}{\gamma(0) \cdot \frac{C(t)}{C(0)} + p} \tag{2.126}$$

Eqn (2.126) is consistent with eqn (2.122) when $p = 0$.

To determine the total number of crosslinks in the system at any time, the recombination reaction must also be considered. Since it seems likely that recombination of a loose crosslink can occur in any segment of the polymer molecule, and since there are usually about one hundred or so segments for every crosslink, the simple assumption of a first-order reaction of ruptured crosslinks seems justifiable. The following kinetic scheme for crosslink reactions is therefore suggested.

$$
\begin{array}{ccccc}
-\,C\,- & & -\,C\,- & & -\,C\,- \\
| & O_2 & | & \text{scission} & | \\
\to Sn & \longrightarrow & Sn-O_2 & \xrightarrow{\quad} & Sm \\
| & & | & & Sm' \\
-\,C\,- & & -\,C\,- & & -\,C\,- \\
& & & & | \\
\text{Structure} & & \text{Structure} & & \text{Structure} \\
A & & B & & C
\end{array}
$$

Recombination

A, B and C are the numbers of structures of types shown in the reaction scheme, k_1 is the velocity constant for oxidation reaction, k_2 that for scission reaction, and k_3 that for recombination. Thus:

$$A + O_2 \xrightarrow{\ k_1\ } B$$

$$B \xrightarrow{\ k_2\ } C$$

$$C \xrightarrow{\ k_3\ } A$$

From the above results and the usual kinetic considerations, the following equations are obtained:

$$\frac{dA}{dt} = k_3 C - k_1 A \tag{2.127}$$

$$\frac{dB}{dt} = k_1 A - k_2 B \tag{2.128}$$

$$\frac{dC}{dt} = k_2 B - k_3 C \tag{2.129}$$

Since A and B are crosslinking structures,

$$C(t) = A + B \tag{2.130}$$

These equations yield the following solutions:

$$\frac{C(t)}{C(0)} = A_1 \left(1 + \frac{k_1}{k_2}\right) - \left(\frac{1}{k_3}\right)(\alpha + k_1) A_2 e^{\alpha t} - \left(\frac{1}{k_3}\right)(\beta + k_1) A_3 e^{\beta t} \tag{2.131}$$

where

$$A_1 = \frac{1 + C_0}{1 + \dfrac{k_1}{k_2} + \dfrac{k_1}{k_3}} \tag{2.132}$$

$$A_2 = \frac{1}{\alpha - \beta}\left\{ k_3 C_0 - (k_1+\beta)a_0 + \beta \cdot \frac{1+C_0}{1 + \dfrac{k_1}{k_2} + \dfrac{k_1}{k_3}} \right\} \tag{2.133}$$

$$A_3 = \frac{1}{\beta - \alpha}\left\{ k_3 C_0 - (k_1+\alpha)a_0 + \alpha \cdot \frac{1+C_0}{1 + \dfrac{k_1}{k_2} + \dfrac{k_1}{k_3}} \right\} \tag{2.134}$$

$$\alpha = -\frac{1}{2}\left\{ (k_1+k_2+k_3) + \sqrt{(k_1+k_2+k_3)^2 - 4(k_1 k_2 + k_2 k_3 + k_3 k_1)} \right\} \tag{2.135}$$

$$\beta = -\frac{1}{2}\left\{ (k_1+k_2+k_3) - \sqrt{(k_1+k_2+k_3)^2 - 4(k_1 k_2 + k_2 k_3 + k_3 k_1)} \right\} \tag{2.136}$$

a_0 is the initial fraction of unreacted crosslinks (in Structure A) and C_0 is the initial fraction of ruptured crosslinks (in Structure C).

If $4(k_1 k_2 + k_2 k_3 + k_3 k_1) > (k_1+k_2+k_3)^2$, the solution is

$$\frac{C(t)}{C(0)} = A_1\left(1 + \frac{k_1}{k_2}\right) - \frac{\cos \omega t}{k_3}\left\{(k_1-\delta)A_2' + \omega A_3'\right\} + \left\{(k_1-\delta)A_3' - \omega A_2'\right\}\sin \omega t \cdot t^{-\delta t} \tag{2.137}$$

where

$$A_2' = a_0 - \frac{1+C_0}{1 + \dfrac{k_1}{k_2} + \dfrac{k_1}{k_3}} \tag{2.128}$$

$$A_3' = (\delta - k_1)a_0 + k_3 C_0 - \frac{(1+C_0)\delta}{1 + \dfrac{k_1}{k_2} + \dfrac{k_1}{k_3}} \tag{2.139}$$

$$2\delta = k_1 + k_2 + k_3 \tag{2.140}$$

$$2\omega = \sqrt{4(k_1 k_2 + k_2 k_3 + k_3 k_1) - (k_1+k_2+k_3)^2} \tag{2.141}$$

IV THE LOCATION OF SCISSIONS ALONG MAIN CHAINS OR AT CROSSLINKAGES

1 Mechanical Measurement

(A) The Case of Cleavage along Network Chains

The relation between the relative stress $f(t)/f(0)$ and the number of scissions along main chains $q_m(t)$ is represented by eqn (2.20), which was previously described in connection with the degradation of crosslinked polymers.

$$\frac{f(t)}{f(0)} = e^{-\frac{q_m(t)}{N(0)}} \tag{2.20}$$

We shall assume that after time t the total number of main chain scissions is $q_{1m}(t)$, $q_{2m}(t)$ and $q_{3m}(t)$, and the relative stress is $f_1(t)/f(0)$, $f_2(t)/f(0)$, and $f_3(t)/f(0)$ for polymers with crosslink densities $N_1(0)$, $N_2(0)$ and $N_3(0)$, respectively. The results in Table 2.5 are obtained.

TABLE 2.5

Mechanism of scission along main chains

	Total number monomers. (per cubic centimeter)	Polymerization degree between two crosslinkages.	Number of network chains. (per cc)	Number of main chain scissions. (per cc)	Relative stress
I	M_0	x_1	$N_1(0)$	$q_{1m}(t)$	$f_1(t)/f(0)$
II	M_0	x_2	$N_2(0)$	$q_{2m}(t)$	$f_2(t)/f(0)$
III	M_0	x_3	$N_3(0)$	$q_{3m}(t)$	$f_3(t)/f(0)$

If $x_1 > x_2 > x_3$, $N_1(0) < N_2(0) < N_3(0)$ is obtained from the relation of $M_0 = xN(0)$ in Table 2.5.

When scission along main chains occurs, the total number of chain scissions $q_m(t)$ after time t is independent of the crosslinking density $N(0)$, therefore,

$$q_{1m}(t) = q_{2m}(t) = q_{3m}(t) \tag{2.142}$$

Accordingly, substituting eqn (2.142) into eqn (2.20),

$$\frac{f_1(t)}{f(0)} < \frac{f_2(t)}{f(0)} < \frac{f_3(t)}{f(0)} \tag{2.143}$$

Eqn (2.142) is illustrated in Fig. 2.43, while eqn (2.143) generates Fig. 2.44.

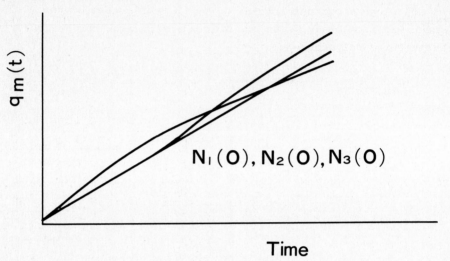

Fig. 2.43. $q_m(t)$ vs. t for the samples having different initial densities N(O).

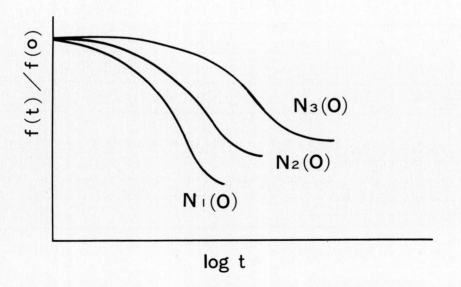

Fig. 2.44. $\frac{f(t)}{f(0)}$ vs. log t for the samples having different initial densities N(O).

(B) The Case of Cleavage at Crosslinkages

We shall assume that, after time t, the total number of scissions at crosslinkages is $q_{1c}(t)$, $q_{2c}(t)$, and $q_{3c}(t)$, and the relative stress is $f_1(t)/f(0)$, $f_2(t)/f(0)$ and $f_3(t)/f(0)$ for the crosslink densities $N_1(0)$, $N_2(0)$ and $N_3(0)$, respectively. See Table 2.6.

TABLE 2.6

Mechanism of scission at crosslinkages

	Total number of monomers (per cc)	Polymerization degree between two cross-linkages	Number of network chains (per cc)	Number of crosslinkages (per cc)	Number of scissions at cross-linkages (per cc)	Relative stress
I	M_0	x_1	$N_1(0)$	$\dfrac{N_1(0)}{2}$	$q_{1c}(t)$	$f_1(t)/f(0)$
II	M_0	x_2	$N_2(0)$	$\dfrac{N_2(0)}{2}$	$q_{2c}(t)$	$f_2(t)/f(0)$
III	M_0	x_3	$N_3(0)$	$\dfrac{N_3(0)}{2}$	$q_{3c}(t)$	$f_3(t)/f(0)$

If $x_1 > x_2 > x_3$, $N_1(0)/2 < N_2(0)/2 < N_3(0)/2$ is obtained from the relation of $M_0 = xN(0)$ because the number of crosslinkages is half that of network chains in an ideal network in Table 2.6.

When scission at crosslinkage occurs under the same conditions, the relation between the number of scissions $q_c(t)$ after time t and the crosslinking density $N(0)$ for three cross-link densities $N_1(0)$, $N_2(0)$ and $N_3(0)$ is given by,

$$\frac{q_{1c}(t)}{\dfrac{N_1(0)}{2}} = \frac{q_{2c}(t)}{\dfrac{N_2(0)}{2}} = \frac{q_{3c}(t)}{\dfrac{N_3(0)}{2}} \qquad (2.144)$$

Suppose the number of crosslink scissions is $q_c(t)$ and the number of network chain scissions, $q_m'(t)$. As two network chains are lost by rupture of one cross-link, the relation between $q_m'(t)$ and $N(0)$ is given by:

$$\frac{q'_{1m}(t)}{N_1(0)} = \frac{q'_{2m}(t)}{N_2(0)} = \frac{q'_{3m}(t)}{N_3(0)} \qquad (2.145)$$

Substituting eqn (2.145) into eqn (2.20) indicates that the relative stress $f(t)/f(0)$ is independent of $N(0)$ when scission at crosslinkage

74

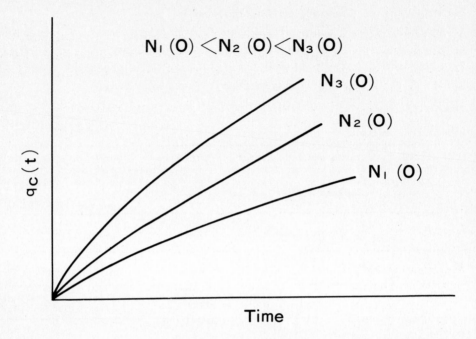

Fig. 2.45. $q_c(t)$ vs. t for the samples having different initial densities N(O).

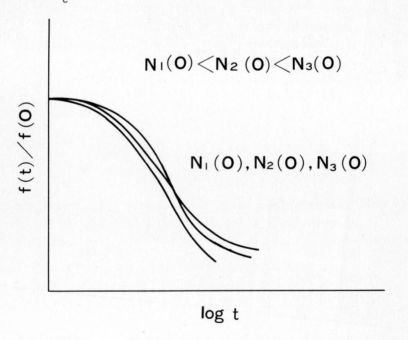

Fig. 2.46. $\frac{f(t)}{f(O)}$ vs. log t for the samples having different initial densities N(O).

occurs.

These results are illustrated in Fig. 2.45 and Fig. 2.46.

In summary, Figs. 2.43 and 2.44 illustrate random scission along network chains, while Figs. 2.45 and 2.46 show the result of scission at crosslinkages.

As an example, oxidative stress-relaxation of radiation-crosslinked natural rubber was carried out (ref. 62). Natural rubber in the form of a thin sheet cast from the latex was crosslinked by exposing it to a 100 GJ mol^{-1} electron beam from a Van de Graaf source. Such samples are eminently suited for chemical stress-relaxation studies because they introduce a minimum of chemicals which may subsequently act as pro-oxidants or antioxidants. The use of chemical cures is greatly complicated for this reason. Various samples were prepared of differing crosslink density, i.e., different values of n(0). The n(0) values for the samples were determined by measuring the stress at 100% elongation and using the equation of state for rubber.

The subsequent stress-relaxation experiments at 80°C and at 130°C are shown in Figs. 2.47 and 2.48, respectively.

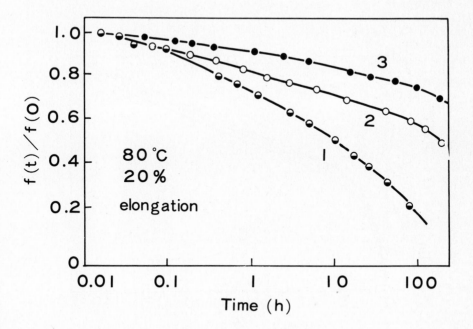

Fig. 2.47. Stress-decay curves at 80°C of radiation-cured natural rubber of different crosslink densities. 1. 2 min, 2. 5 min, 3. 10 min. (Time of exposure to 100 GJ mol^{-1} electrons at 1 μA)

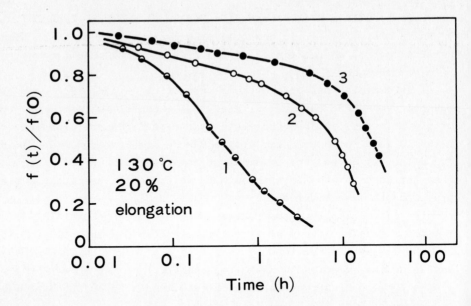

Fig. 2.48. Stress-decay curves at 130°C of radiation-cured natural rubber of diffe
crosslink densities. 1. 2 min, 2. 5 min, 3. 10 min. (Time of exposure to 100 GJ m
electrons at 1 μA)

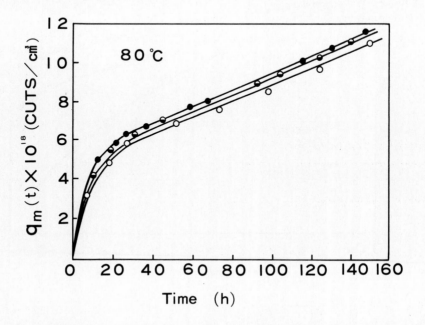

Fig. 2.49. $q_m(t)$ vs. t for radiation-cured natural rubber at 80°C.

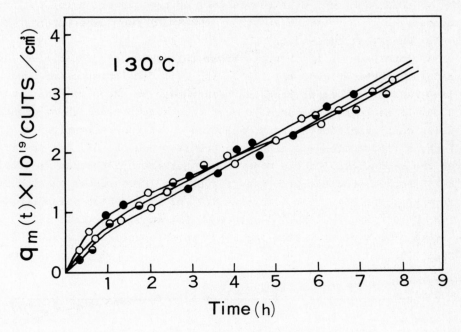

Fig. 2.50. q_m(t) vs. t for radiation-cured natural rubber at 130°C.

The same data are shown plotted in the form of q_m(t) versus time in
Figs. 2.49 and 2.50 where q_m(t) was obtained from eqn (2.21).

The results clearly follow the pattern discussed under section
2.4.1., the case of random scission of the links of the network
chains as opposed to scission at the cross-links. Explicitly, q_m(t)
is seen to be independent of N(O) but the f(t)/f(O) versus t curves
clearly depend on N(O), as predicted for random scission of main
chains. This result has also been found for natural rubber crosslinked
by azodicarboxylates.

The scissions in unvulcanized natural rubber exposed to air at
80°C and at 130°C were measured by following the change in number-
average molecular weight with time. The sample used was in the form
of a thin film of rubber, and the molecular weight as a function of
time was determined by withdrawing small samples and measuring the
intrinsic viscosity. The quantity Q(t), the moles of cleavage per
cubic centimeter of rubber, was determined by the equation.

$$\varrho(t) = d \left\{ \frac{1}{\overline{M}_n(t)} - \frac{1}{\overline{M}_n(0)} \right\}$$ (2.146)

In this equation, d is the density of the rubber, $\bar{M}_n(t)$ the numbe
average molecular weight at time t and $\bar{M}_n(0)$ the number-average mole
cular weight at time zero.

A comparison of $q_m(t)$ from stress-relaxation of radiation-cured
natural rubber at 80°C and 130°C with $Q(t)$ from scission of unvulcan
ized natural rubber at the same temperature is shown in Figs. 2.49 a
2.50. It is seen that there is good agreement between these quantiti
which must mean that scission measured by stress relaxation is, in
fact, a random cleavage along the polyisoprene chains, not specifica
at the crosslinks. On the other hand, tetramethyl thiuram di-sulfide
(TMTD)-cured ethylene-propylene terpolymer (EPT) rubber behaved dif-
ferently. The crosslink densities $N(0)$ of this rubber were 0.62×10
1.05×10^{-4} and 1.32×10^{-4} mol cm^{-3}. The chemical stress-relaxation
curves of $f(t)/f(0)$ versus log t for three specimens are shown in
Fig. 2.51.

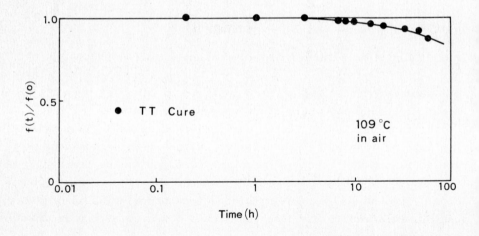

Fig. 2.51. Pure chemical stress-relaxation of TMTD-cured EPT.

As seen from this figure, the curves were found to be independent
of $N(0)$ (ref. 63). So the $q_m(t)$ versus time curves were seen to be
dependent on $N(0)$ from the above reason. Therefore, the results
clearly follow the pattern discussed under section 2.4.1.(B), the
case of scission at crosslinks.

The structure of monosulfide -S-, and/or disulfide -S$_2$- linkage
in TMTD-cured EPT was found to follow the above pattern.

2 Modified Mechanical and Swelling Measurement (ref. 64)

There is the another method for the estimation of the number and
the site of chain scissions, based upon the determination of equilibrium degree of swelling which was developed by Horikx (ref. 65) and
Charlesby (ref. 66).

Stress-decay is not necessarily caused by oxidative scission alone,
but also by the relaxation of trapped entanglements, exchange reactions, etc. The criteria described above cannot be applied to stress-
decay other than oxidative scission. Second, it is not easy to prepare
rubber samples crosslinked to different degrees but otherwise identical.
Sulfur-cured rubber vulcanizates contain not only cyclic and crosslinking sulfur groupings but also accelerator fragments and unreacted
sulfur, these appear to have a considerable effect on the overall
rate of oxidative scission. Some of these entities are directly
attached to the network chain and cannot be removed by solvent extraction. It seems likely that the number of such bound groupings should
increase with increasing network chain density and that may have an
appreciable influences upon the rate of chain scission. The measurement
of the equilibrium degree of swelling, however, is too time-consuming
for practical use, since it requires a long time to elapse until the
equilibrium swelling state is reached.

In order to overcome the difficulties encountered in the above-
described methods, a simple method for the discrimination of the
chain scission mechanism was proposed, as follows (ref. 64).

Combining Horikx's equations with the mechanical ones, we have,
(1) Random scission along the main chain

$$\frac{f(t)}{f(0)} = \left[\frac{1-S(t)^{1/2}}{1-S(0)^{1/2}} \right]^2 \qquad (2.147)$$

(2) Scission of crosslinks

$$\frac{f(t)}{f(0)} = \frac{\gamma(t)\{1-S(t)^{1/2}\}^2}{\gamma(0)\{1-S(0)^{1/2}\}^2} \qquad (2.148)$$

(3) Selective scission near crosslink sites

$$\frac{f(t)}{f(0)} = \left[\frac{1-\{2S(t)\}^{1/2}}{1-S(0)^{1/2}} \right]^2 \qquad (2.149)$$

where, $S(0)$ and $S(t)$ are sol fractions at time zero and t, and $\gamma(0)$ and $\gamma(t)$ are corresponding values of the average number of crosslink per primary polymer chain.

Provided that the distribution of primary chains is random (ref. we may apply Charlesby's relationship, which is given by (ref. 66).

$$1/\gamma = S+S^{1/2} \tag{2.150}$$

Then, eqn (2.148) is reduced to,

$$\frac{f(t)}{f(0)} = \frac{\{S(0)+S(0)^{1/2}\}\{1-S(t)^{1/2}\}^2}{\{1-S(0)^{1/2}\}^2\{S(t)+S(t)^{1/2}\}} \tag{2.151}$$

On applying eqns (2.147), (2.149) and (2.151), the scission mecha of an arbitrary rubber vulcanizate is easily discerned. Thus, the so fractions of aged samples are determined at appropriate time interva and the right hand side values of eqns (2.147), (2.149) and (2.151) are calculated as functions of time. Then, these values are compared with a stress-relaxation curve to distinguish which values correspon with the curve.

Some attention should be paid to the method of measuring stress-relaxation. The fundamental assumption is that the crosslinking reaction does not affect the stress value obtained by "continuous stress relaxation", in which the sample is held at constant length (68). On the other hand, contributions from scission and crosslinking are both reflected in equilibrium swelling measurements. Hence, "intermittent stresses" must be preferred as the relative stress values. In this method, a relaxed sample is rapidly stretched to a constant elongation at appropriate time intervals, and the stress is quickly measured (ref. 68).

The raw rubbers used in this investigation were natural rubber (NR), synthetic cis-1,4-polyisoprene (cis-PI) and synthetic cis-1,4-polybutadiene (cis-PB). The cis-PI (Nipol 2200) had 98.5% cis-1,4-units. The microstructure of the cis-PB (Ubepol-BR 110) was cis-1,4 98%, $trans$ 1% and vinyl 1%.

Compounding and curing conditions are summarized in Table 2.7 for dicumyl peroxide (DCP) curing and in Table 2.8 for sulfur (Sx) and tetramethyl thiuram disulfide (TMTD) curing, respectively. All the samples were extracted by hot acetone or by cold benzene for 24~48 h Continuous and intermittent stresses were measured by using a

TABLE 2.7

Compounding and curing conditions for samples cured with dicumyl peroxide

	NR, *cis*-PI	*cis*-PB
Rubber (wt. %)	97.1	98.0
Dicumyl Peroxide (wt. %)	2.9	2.0
Curing Temp. (°C)	145	130
Curing Time (min,)	30	30

TABLE 2.8

Compounding and curing conditions for samples cured with sulfur and tetramethyl thiuram disulfide

	NR	cis-PI	cis-PB	cis-PI
Rubber (wt. %)	91.7	90.9	91.3	93.5
ZnO (wt. %)	4.6	4.5	4.6	4.7
Stearic Acid (wt. %)	0.9	0.9	0.9	
Sulfur (wt. %)	1.8	1.8	1.8	
Dibenzothiazyl Disulfide (wt. %)	0.9		0.9	
Diphenylguanidine (wt. %)		0.9	0.5	
Tetramethyl Thiuram Disulfide (wt. %)				1.8
Curing Temp. (°C)	143	145	145	145
Curing Time (min.)	30	15	11	20

conventional stress relaxometer, the measurements being started after 20 min pre-heating. The stress value at 36 s after extension was taken as the initial stress.

Sol fractions were determined as follows. An aged sample was immersed in benzene in a glass tube for 24~30 h. After the complete dissolution of the soluble part, the remaining gel was weighed after drying. The sol fraction is calculated from the weight loss during degradation. In order to adopt the same initial condition as for the mechanical measurements, the sample pre-heated for 20 min was taken as the undegraded specimen.

For highly degradable samples, a special cylindrical holder which consists of 100 mesh stainless steel net was designed to avoid adhesion to the glass tube. The sol fraction was determined by weighing the sample together with the holder.

In order to confirm the applicability of the above-mentioned method, the oxidative degradation of DCP-cured NR (NR-DCP) was investigated. It has been well established that the scission occurs randomly along the main chain for this vulcanizate (refs. 65,68), moreover it is known that no crosslinking reaction occurs in this case.

As shown in Fig. 2.52, the continuous and intermittent stress-relaxa-

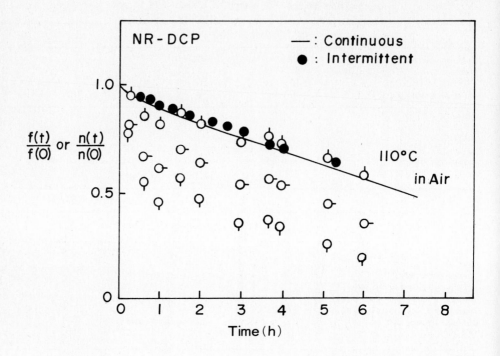

Fig. 2.52. Change in effective network chain-density with time at 110°C for DCP-cured natural rubber (NR-DCP). Solid curve and filled circles represent continuous and intermittent stress-relaxation data, respectively. Open circles with pips denote the values calculated from sol fraction by assuming random chain-scission along the main chain (pips up), selective scission near crosslinks (pips right), and scission at crosslink sites (pips down). See eqns (2.147), (2.151) and (2.149) for details.

tion curves coincide with each other, indicating the absence of crosslinking reactions. The right hand side values of eqn (2.147) are calculated from the sol fraction at each time and are also plotted in this figure, giving values in good agreement with those obtained from stress-relaxation. On the other hand, the values calculated from eqns (2.149) and (2.151) deviate markedly from the experimental stress-relaxation curve.

A similar experiment was carried out for sulfur-cured natural rubber (NR-Sx), as shown in Fig. 2.53. The values obtained from the sol fraction measurement, assuming random scission along the main chain, again agree well with those obtained from intermittent stress data, but they do not coincide with the continuous stress-decay curve. As discussed previously, disagreement with the continuous stress-decay curve

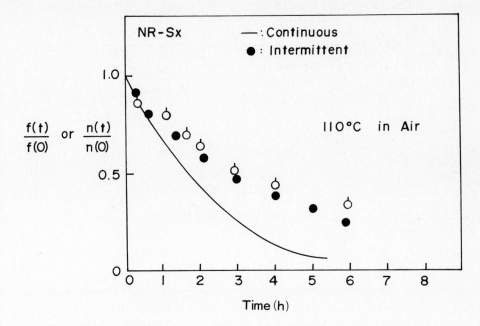

Fig. 2.53. Change in effective network chain-density with time at 110°C for sulfur-cured natural rubber (NR-Sx). The symbols are the same as in Fig. 2.52.

may be ascribed to crosslinking during the course of degradation.

These results clearly show that the oxidative chain-scission of natural rubber vulcanizates occurs randomly along the main chain, irrespective of the crosslink structure. This is consistent with the conclusions of Tobolsky (ref. 68) and Horikx (ref. 65), indicating the applicability of this simple method of discrimination between possibe chain-scission mechanisms.

In order to apply our method to other vulcanized rubber systems, the oxidative chain-scission of cis-1,4-polyisoprene and cis-1,4-polybutadiene vulcanizates was investigated.

It is generally accepted that the oxidative degradation of a well extracted sample of DCP-cured cis-1,4-polyisoprene (PI-DCP) is much faster than that of NR-DCP. The stress-decay curve of PI-DCP has an autocatalytic character, while the stress decays steadily in NR-DCP showing the typical feature of inhibited oxidation (Fig. 2.54). The rather slow rate of oxidative stress-decay in NR-DCP is generally ascribed to the inhibitory effect of the impurities in NR which cannot be extracted by organic solvents and, hence, its decay curve may not be regarded as characteristic of the primary chemical structure of the main chain. Therefore, the oxidative degradation of well extracted

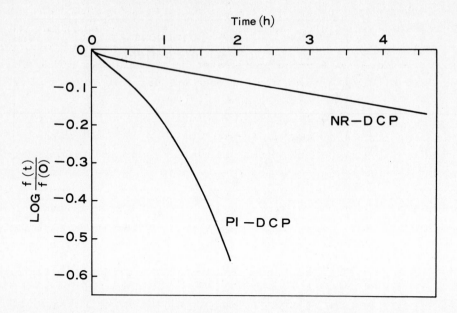

Fig. 2.54. Relationships between logarithmic relative stress and time in continuous stress-decay experiments for DCP-cured natural rubber (NR-DCP) and *cis*-1,4-poly-isoprene (PI-DCP) at 110°C. The curve obtained for NR-DCP shows the characteristic feature of inhibited oxidation.

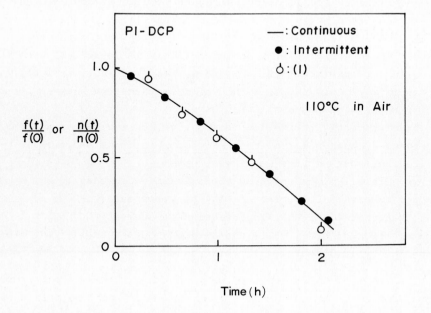

Fig. 2.55. Change in effective network chain-density with time at 110°C for DCP-cured *cis*-1,4-polyisoprene (PI-DCP). The symbols are the same as in Fig. 2.52.

PI-DCP should be studied in order to clarify the inherent behavior
of polyisoprene. Results show (Fig. 2.55) that both continuous and
intermittent stress-decay curves follow the equation based on random
scission. Hence, in spite of the different shape of their stress-
decay curve, the chain-scission of both DCP vulcanizates has been
proven to take place along the main chain.

The rate of stress-decay of sulfur-cured cis-1,4-polyisoprene
(PI-Sx) is similar to that of NR-Sx. Thus, the chain-scission
mechanisms of these vulcanizates are presumed to be identical. On
the other hand, there remains some ambiguity with respect to the
chain-scission mechanism of TMTD-cured cis-1,4-polyisoprene (PI-TMTD).
In order to resolve this point, we have studied the oxidative degradation of PI-TMTD (Fig. 2.56). In this case also, the experimental
results clearly show that scission occurs along the main chain.

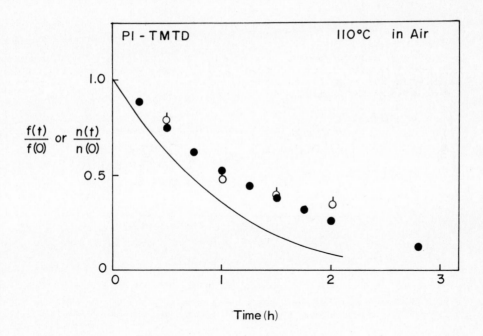

Fig. 2.56. Change in effective network chain-density with time at 110°C for TMTD-
cured cis-1,4-polyisoprene (PI-TMTD). The symbols are the same as in Fig. 2.52.

Thus, it is seen that the stress-decay of all the cis-1,4-polyisoprene
vulcanizates is caused by random chain-scission along the main chain,
irrespective of the differences in rate and the shape of the stress-
decay curves.

86

.Little attention has been given to the thermal oxidation behavior
of cis-1,4-polybutadiene vulcanizates. In contrast to polyisoprene
rubbers, crosslinking reactions become significant for this polymer
(refs. 69,70). In order to obtain information about the chain-scissi
mechanism for cis-1,4-polybutadiene vulcanizates, our method has bee
applied.

The decay of intermittent stress for DCP-cured cis-1,4-polybutadi
(PB-DCP) was compared with the values calculated from eqns (2.147),
(2.149) and (2.151) using sol fraction data (Fig. 2.57). The figure
suggests strongly that the scission occurs selectively at the main
chain crosslink sites.

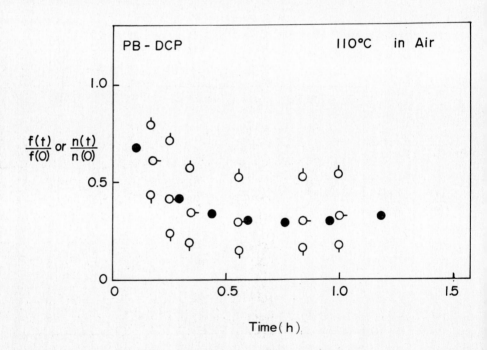

Fig. 2.57.Change in effective network chain-density with time at 110°C for DCP-
cured cis-1,4-polybutadiene (PB-DCP). The symbols are the same as in Fig. 2.52.

Furthermore, it should be noted that appreciable crosslinking takes
place during the course of degradation. In the later stages, crosslink formation seems to be in equilibrium with chain-scission and th
number of effective network chains remains nearly constant.

These results are quite different from those obtained for PI-DCP
(Fig. 2.55). In order to clarify the difference between the degradat
behavior of PI-DCP and PB-DCP, the numbers of chain-scissions are

compared for these vulcanizates at the same temperature (Fig. 2.58).
The stress-decay was much faster for PB-DCP than for PI-DCP.

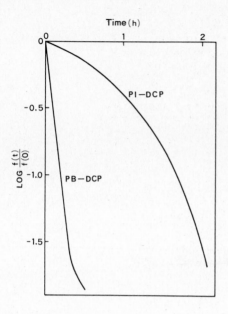

Fig. 2.58.Plots of logarithmic relative stress against time in continuous stress-
decay experiments for DCP-cured cis-1,4-polyisoprene (PI-DCP) and cis-1,4-poly-
butadiene (PB-DCP).

All of these results point to the presence of weak points near
crosslink sites in PB-DCP which are highly reactive to oxidative
chain-scission.

Fig. 2.59 shows the result for sulfur-cured cis-1,4-polybutadiene
(PB-Sx). Although some deviations due to crosslinking reaction were
observed in the later stages, the figure indicates that random chain-
scission is predominant in this system. The rate of chain-scission
for PB-Sx is compared with that for PI-Sx in Fig. 2.60.

Oxidative chain-scission for these rubbers appears to proceed
steadily without showing the concave curvature characteristic of
autocatalytic process. Moreover, rates of chain-scission are the same
within experimental error. This result is consistent with the oxygen
absorption studies of Gonda et al. (ref. 71). In view of the similarity
in chemical structure of PI and PB, it appears likely that the same
type of radical species is responsible for the oxidative degradation
of PI-Sx and PB-Sx, and that the methyl group of PI does not play an
important role in the oxidation of sulfur vulcanizates.

88

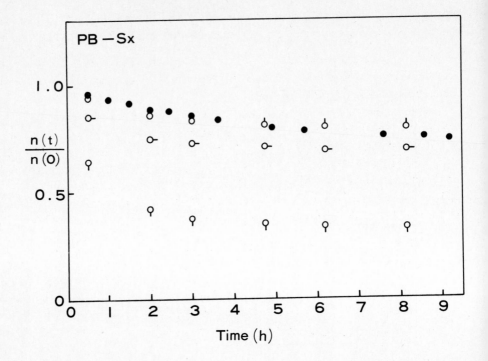

Fig. 2.59. Change in effective network chain-density with time at 110°C for sulfur-cured *cis*-1,4-polybutadiene (PB-Sx). The symbols are the same as in Fig. 2.52.

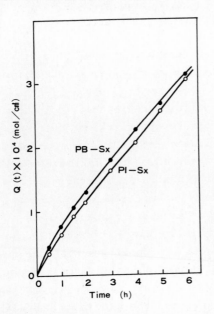

Fig. 2.60. Plots of the number of chain-scissions against time for sulfur-cured 1,4-polyisoprene (PI-Sx) and *cis*-1,4-polybutadiene (PB-Sx).

Comparing the degradation behavior of sulfur-cured vulcanizates with those of DCP-cured ones, it may be concluded that the unusual features observed with PB-DCP are not characteristic of the *cis*-1,4-polybutadiene chain, but are to be ascribed to the specific crosslink structure formed by DCP curing.

Let us consider the difference in the crosslinking reactions of PI and PB using DCP as curing agent. In these rubbers, the initiation reaction may be hydrogen abstraction from the C-H bond adjacent to a double bond, leaving a polyisoprenyl or polybutadienyl radical. Crosslinks are mainly formed by the recombination of these radicals. Now, these radicals are stabilized by allyl resonance:

(I)

(II)
(polyisoprenyl radical)

(III)

(IV)
(polybutadienyl radical)

It has been suggested that the polybutadienyl radical is more reactive than the polyisoprenyl radical, and that attack on double bonds by this radical is an alternative route to recombination for the formation of crosslinks (ref. 72).

This reaction is represented by,

In the case of polybutadiene, it is evident that only crosslinks with tertiary C-H bonds are formed, whether they are generated by recombination or attack of polybutadienyl radicals on double bonds. One might expect that the reactivity of the tertiary hydrogen atoms on the crosslinks are much higher than that of the secondary hydrogen atoms on the main chain. Therefore, the hydrogen atoms of the cross-link points may be selectively abstracted, leading to chain-scission

90

near crosslink sites.

In the case of PI, however, structures (I) and(II) give crosslink
with tertiary and quarternary carbon, respectively. The crosslinks
with quarternary carbon are regarded as quite stable to oxidation,
since they have no labile C-H bonds but crosslinks with tertiary
hydrogen atoms may be quite susceptible to hydrogen abstraction.

The reactivity toward oxidation of PI-DCP and PB-DCP should be in
the same order, provided structure (I) is predominant in the oxidati
of PI-DCP. The observed large differences in the rate of oxidation
for PI-DCP and PB-DCP, therefore, can only be explained by consideri
the exclusive formation of quarternary carbon, that is, the formatio
of structure (II). In this case, the rate of oxidation of PI-DCP wil
be determined by the reactivity of secondary C-H bounds in the main
chain, which are more stable than tertiary C-H bonds. Although we
have no chemical evidence for the exclusive formation of structure
(II), the observed high rate of oxidation and selective chain-scissio
near crosslinks in PB-DCP vulcanizate may be well explained by the
above hypothesis.

REFERENCES

1 A.M. Bueche, J. Chem. Phys., 21(1953)614.
2 J.P. Berry and W.F. Watson, J. Polym. Sci., 18(1955)201.
3 H.M. James and E. Guth, J. Chem. Phys., 11(1943)470; W. Kuhn, Kolloid-Z., 76
  (1936)258; L.R.G. Treloar, Trans. Faraday Soc., 39(1943)241; P.J. Flory and J.
  Rehner, J. Chem. Phys., 11(1943)512.
4 J.P. Berry and W.F. Watson, J. Polym. Sci., 18(1955)201.
5 H.Yu, Polym. Lett., 2(1964)631.
6 A.V. Tabolsky, D.J. Metz and R.B. Mesrobian, J. Am. Chem. Soc., 72(1950)1942.
7 J.G. Curro, J. Polym. Sci., Polymer Physics Ed., 14(1976)177.
8 A.V. Tobolsky, Properties and Structure of Polymers, John Wiley, New York,
  1960, p.336.
9 H. Mark and A.V. Tobolsky, Physical Chemistry of High Polymeric Systems, John
  Wiley, New York, 1950, p.353; A.V. Tobolsky, Properties and Structure of Polym
  John Wiley, New York, 1960, p.234.
10 A.V. Tobolsky, I.B. Prettyman and J.H. Dillon, J. Appl. Phys., 15(1944)324.
11 J. Scanlan, J. Polym. Sci., 43(1960)501.
12 A.V. Tobolsky, P.M. Norling, N. Frick and H.Yu, J. Am. Chem. Soc., 86(1964)392
13 A.V. Tobolsky, Polym. Lett., 2(1964)637.
14 A.M. Bueche, J. Chem. Phys., 21(1953)614.
15 M.T. Shaw, J. Chem. Phys., 54(1971)2172.
16 A. Kaeriyama, K. Ono and K. Murakami, Makromol. Chem., 175(1974)3011.
17 J.R. Shelton and D.N. Vincent, J. Am. Chem. Soc., 85(1963)2433.
18 M.B. Neiman (Editor), Aging and Stabilization of Polymers, Consultants Bureau,
   Plenum Publishing Corporation, New York, 1965.
19 J.R. Shelton and D.N. Vincent, J. Am. Chem. Soc., 85(1963)2433.
20 G.W. Blum, J. R. Shelton and H. Winn, Ind. Eng. Chem., 43(1951)464.
21 J.P. Berry, J. Polym. Sci., 21(1956)505.
22 K. Ono, A. Kaeriyama and K. Murakami, J. Polym. Sci., Polym. Chem. Ed., 13
   (1975)1209.
23 K. Ono, A. Kaeriyama and K. Murakami, ibid., 16(1978)1575.
24 J. Crank, The Mathematics of Diffusion, Oxford Univ. Press, London, 1975.

25 A.V. Tobolsky, D.J. Metz and R.B. Mesrobian, J. Am. Chem. Soc., 72(1950)1942.
26 G.J. Amerongen, J. Appl. Phys., 17(1946)972.
27 N.C. Billingham and J.T. Walker, J. Polym. Sci., Polymer Chem. Ed., 13(1975)1209.
28 P.V. Danckwerts, Trans. Faraday Soc., 47(1951)1014.
29 M. Eigen and L. DeMayer, Technique of Organic Chemistry, A. Weissberger et. al.,
   Ed., Vol.8, Part II, Interscience, 1963.
30 R.M. Barrer, Nature, Lond. 140(1937)106.
31 G.J. Amerongen, J. Polym. Sci., 5(1950)307.
32 S.W. Benson, J. Am. Chem. Soc., 87(1965)972.
33 S.M. Katz, E.T. Kubu and J.H. Wakelin, Text. Res. J., 20(1950)754.
34 C.E. Rease and H. Eyring, ibid., 20(1950)743.
35 D.R. Paul, J. Polym. Sci., A-2, 7(1969)1811.
36 J.H. Petropoulos, ibid., 8(1970)1797.
37 J.A. Tshudy and C. von Frankenberg, J. Polym. Sci., Polymer Phys. Ed., 11(1973)
   2027.
38 H.H.G. Jellinek, J. Polym. Sci., Polymer Chem. Ed., 14(1976)1249.
39 W. Kuhn, Kolloid-Z., 76(1936)258; F.T. Wall, J. Chem. Phys., 11(1943)67; H.M.
   James and E. Guth, J. Chem. Phys., 11(1943)470; L.R.G. Treloar, Trans. Faraday
   Soc., 39(1943)241.
40 P.J. Flory, Principles of Polymer Chemistry, Cornell Univ. Press, Ithaca, N.Y.,
   1953, ch. 13..
41 G. Kraus, Rubber World, 135(1956)67.
42 P.J. Flory and J. Rehner, J. Chem. Phys., 11(1943)512.
43 L. Mullins, J. Appl. Polym. Sci., 2(1959)1.
44 K. Murakami and G.H. Hisue, Polym. Lett., 10(1972)253.
45 K. Murakami and S. Tamura, unpublished results.
46 K. Murakami and S. Takasugi, Polym. J., 7(1975)588.
47 T. Smith, Paper presented at the memorial meeting of Prof. Tobolsky, Princeton,
   October (1973).
48 S. Tamura and K. Murakami, Polym., 14(1973)570; Rubber Chem. Technol., 48(1975)
   141.
49 K. Murakami and S. Tamura, J. Polym. Sci., B, 11(1973)317.
50 T. Kusano, S. Tamura and K. Murakami, J. Polym. Sci., Symposium No.46(1974)251.
51 K. Murakami and S. Tamura, J. Polym. Sci., B, 11(1973)529.
52 Y. Takahashi and A.V. Tobolsky, Polym. J., 2(1971)457.
53 S. Tamura and K. Murakami, J. Appl. Polym. Sci., 16(1972)1149.
54 K. Murakami and G.H. Hsiue, J. Appl. Polym. Sci., 16(1972)2249.
55 M.D. Stern and A.V. Tobolsky, J. Chem. Phys., 14(1946)93.
56 M. Mochulsky and A.V. Tobolsky, Ind. Eng. Chem., 40(1948)2155.
57 D.H. Johnson, J.R. McLoughlin and A.V. Tobolsky, J. Phys. Chem., 58(1954)1073.
58 R.C. Osthoff, A.M. Bueche and W.I. Grubb, J. Am. Chem. Soc., 76(1954)4659.
59 A.V. Tobolsky, Polym. Lett. 2(1964)823.
60 S. Baxter and H.A. Vodden, Polym. , 4(1963)145.
61 T. Migita and K. Murakami, Chemorheology of Polymers, (Japanese Ed.). Asakura
   Shoten, Japan, 1968, p.90.
62 A.V. Tobolsky, Properties and Structure of Polymers, John Wiley, New York, 1960,
   p.236.
63 K. Murakami and S. Tamura, Polym. J., 2(1971)330; T. Kusano, S. Tamura and K.
   Murakami, J. Polym. Sci., Symposium No.46(1974)254.
64 A. Kaeriyama, K. Ono and K. Murakami, J. Polym. Sci., 16(1978) 2063.
65 M.M. Horikx, J. Polym. Sci., 19(1956)445.
66 A. Charlesby, J. Polym. Sci., 11(1953)513.
67 K.W. Scott, J. Polym. Sci., 58(1962)517.
68 A.V. Tobolsky, Structure and Properties of Polymers, John Wiley, New York, 1960.
69 G. Scott, Atmospheric Oxidation and Antioxidants, Elsevier, London, 1965.
70 E.M. Bevilacqua, J. Polym. Sci., C24(1968)285.
71 K. Gonda, J. Turugi and N. Murata, Nippon Gomu Kyokaishi (Journal of the
   Society of Rubber Industry, Japan), 40(1967)572.
72 B.M.E. van der Hoff, Ind. Eng. Chem., 2(1963)273.
73 R.T. Conley (Editor), A.V. Tobolsky, A.M. Kotliar and T.C.P. Lee, Thermal
   Stability of Polymers, Volume 1, Marcel Dekker Inc., New York, 1970, p.104.

92

CHAPTER 3

MULTI-DEGRADATION OF CROSSLINKED POLYMERS

I CLEAVAGE AT CROSSLINKS AND ALONG MAIN CHAINS

In the case when scissions of both main chain and crosslinks occur
at the same time, the relationship between the change of chemical
structure and the properties of the polymer network chain have still
not been made clear. Therefore, in order to investigate the degree
of agreement between our theoretical relation for both scissions and
the experimental results (ref. 1), we carried out further studies. A
TMTD-cured natural rubber was prepared as a model polymer in which
both scissions of main chain and crosslinkage could occur at the same
time. The initial crosslink density of this rubber is n"(0). The model
polymer TMTD-cured EPT (Ethylene-propylene terpolymer) in which only
crosslink scission occurs under the given conditions, has an initial
crosslink density of n(0), and the number of crosslink scissions $q_c(t)$
is represented by eqn (2.110).

$$q_c(t) = C(0)(1-e^{-kt}) \qquad (2.110)$$

The other model polymer, irradiation-cured natural rubber, in which
only main chain scission occurs, has an initial crosslink density
n'(0) and the number of main chain scissions $q_m(t)$ is represented by
eqn (2.27).

$$q_m(t) = -n'(0)\ln \frac{f(t)}{f(0)} \qquad (2.27)$$

In the case of a TMTD-cured natural rubber, with simultaneous
scissions of main chain and crosslinks, if we consider the chemical
stress-relaxation of the above three samples under the same condition
(i.e., in air at 109°C), we find

$$Q(t) = q_m(t) + \frac{n''(0)}{n(0)} q_c(t)$$

$$= -n'(0)\ln \frac{f(t)}{f(0)} + \frac{n''(0)}{n(0)} \cdot C(0) \cdot (1-e^{-kt})$$

$$= -n'(0)\ln \frac{f(t)}{f(0)} + \frac{n''(0)}{2} \cdot (1-e^{-kt}) \qquad (3.1)$$

Here, $Q(t)$ is the total number of scissions of both main chains and crosslink sites for a TMTD-cured natural rubber. The value of $Q(t)$ of this natural rubber under this condition can be determined from the calculation of eqn (3.1) using the experimental results of the two model samples, TMTD-cured EPT and irradiation-cured natural rubber. Furthermore, the following equation was derived for the case in which scissions of main chain and crosslinkage both occur at the same time (ref. 1):

$$\frac{f(t)}{f(0)} = 1 - \frac{4k(t)\cdot x}{k(t)+2x} \cdot \frac{Q(t)}{M_o} \cdot \exp\left\{-\frac{2x}{k(t)+2x} \cdot \frac{Q(t)}{M_o}\right\}$$

$$+ \frac{2k(t)\cdot x}{k(t)+2x} \cdot \frac{Q(t)}{M_o} \cdot \exp\left\{-\frac{4x^2}{k(t)+2x} \cdot \frac{Q(t)}{M_o}\right\} \qquad (3.2)$$

where

$$k(t) = \frac{q_c(t)/C(0)}{q_m(t)/M_o}$$

and $\qquad (3.3)$

$$M_o = x \cdot n(0)$$

M_o is the total number of monomer units per cubic centimeter of crosslinked polymer, and x is the average number of monomer units between two crosslinkages. In eqn (3.2), for a TMTD-cured natural rubber, the experimental values of $Q(t)$, $q_c(t)$, M_o and $k(t)$ are already known and are tabulated in Table 3.1. The relation between $f(t)/f(0)$ and t of this rubber is predicted in Fig. 3.1 as the calculated curve. From Fig. 3.1, the curve observed usually falls below the calculated one, but there is a little difference between the curves. The main reason for the discrepancy is considered to be based upon the small change of the strength of the partial main chains near the cross-linkages, and this has still not been elucidated quantitatively. In more detail, the chemical structure of crosslinkages in a TMTD-cured natural rubber appears to have the effect of making the partial main chains near the crosslinkages less subject to oxidative scission.

TABLE 3.1

Experimental values of $q_c(t)$, $q_m(t)$ and $Q(t)$ and calculated values of $f(t)/f(0)$ obtained from eqn (3.2)

Time(h)	$q_c(t) \times 10^{-7}$ (mol/ml)	$q_m(t) \times 10^{-8}$ (mol/ml)	$Q(t) \times 10^{-8}$ (mol/ml)	$f(t)/f(0)$
0	0	0	0	1.000
0.5	0.05	4.7	4.7	0.963
1	0.16	8.0	8.0	0.937
1.5	0.31	9.8	9.8	0.925
2	0.62	11.7	11.8	0.909
3	1.24	14.8	14.9	0.888
4	1.86	18.2	18.4	0.863
5	3.09	21.6	21.9	0.839
6	4.02	25.0	25.4	0.817

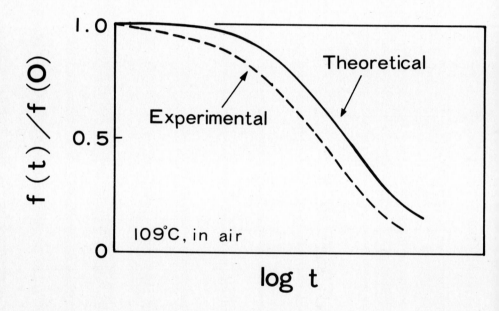

Fig. 3.1. Calculated and observed stress-relaxation curves of TT-cured natural rubber.

In the case of degradation of irradiation-cured natural rubber, when scission occurs only along the main chains in crosslinked polyme eqn (3.4) is obtained and is illustrated in Fig. 3.2.

$$q_m(t) = x \cdot n(0) \cdot k_1 \cdot t \qquad (3.4)$$

Here, x, $n(0)$, and k_1 are, respectively, the average number of monomer units between two crosslinkages, the initial crosslinking

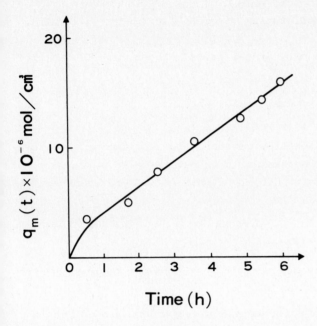

Fig. 3.2. Relationship between $q_m(t)$ and time at 110°C in air for irradiation cured natural rubber vulcanizates.

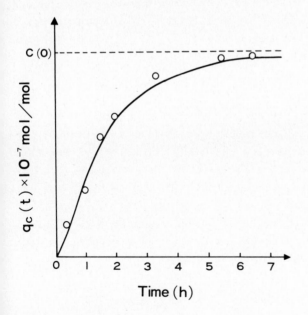

Fig. 3.3. Relationship between $q_c(t)$ and time at 110°C in air for TMTD-cured EPT.

96

density, and the proportionality constant for the scission along main chains.

When scission occurs only at the crosslinkages, eqn (3.5) results as described above

$$q_c(t) = C(0) \cdot (1-e^{-k_2 t})$$

(3.5)

Here, k_2 is the proportionality constant for scission at the cross-links. As an example of a chemical stress-relaxation curve, Fig. 3.3 shows scission occurring at the crosslinks of TMTD-cured EPT at 110°C in air.

From eqns (3.4), (3.5), and (3.3), eqn (3.6) was derived:

$$k(t) = \frac{1-e^{-K \cdot k_1 \cdot t}}{k_1 t}$$

(3.6)

Here, $K = k_2/k_1$, which is the scission ratio for main chains in comparison with crosslinks.

In the case of degradation of irradiation-cured natural rubber in air at 110°C, as shown in Fig. 3.2, k_1 was found to be 2.47×10^{-4} s^{-1} in eqn (3.4) and that of TMTD-cured ethylene-propylene terpolymer, as shown in Fig. 3.3, k_2 was found to be 1.12×10^{-3} s^{-1}. Accordingly, for TMTD-cured natural rubber in air at 110°C,

$$K = \frac{k_2}{k_1} = \frac{1.12 \times 10^{-3}}{2.47 \times 10^{-4}} = 4.53$$

a result which it should be interesting to confirm experimentally, if there were any way of measuring the values of k_1 and k_2 for this vulcanizate under the conditions employed.

The theoretical relations between $f(t)/f(0)$ and $\log t$ in eqn (3.2) are depicted as functions of k_1 and k_2 in Figs. 3.4, 3.5 and 3.6.

From the values of k_1 and k_2, $e^{-k_2 t} \approx 1-k_2 t$ will be established in the initial stages of the observations. Under these conditions, eqn (3.6) will transform as follows.

$$k(t) = \frac{1-e^{-K \cdot k_1 t}}{k_1 t} = \frac{k_2}{k_1} = K$$

(3.7)

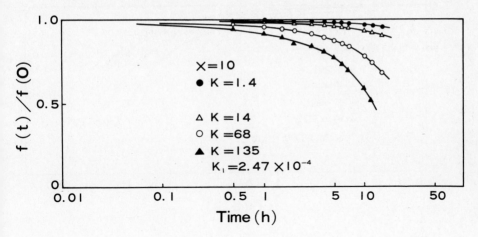

Fig. 3.4. Theoretical relationships between $f(t)/f(0)$ and $\log t$ in eqn (3.2) as the function of k_1 and k_2 when $x=10$.

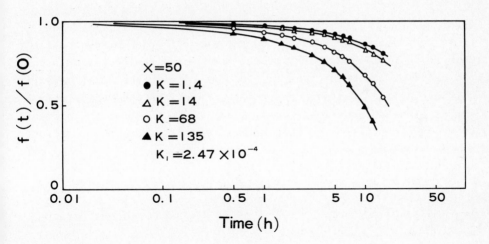

Fig. 3.5. Theoretical relationships between $f(t)/f(0)$ and $\log t$ in eqn (3.2) as the function of k_1 and k_2 when $x=50$.

Furthermore, eqn (3.2) becomes:

$$\frac{f(t)}{f(0)} = (1-2K \cdot k_1 t) \cdot e^{-x \cdot k_1 t} + K \cdot k_1 t e^{-2x \cdot k_1 t} \tag{3.8}$$

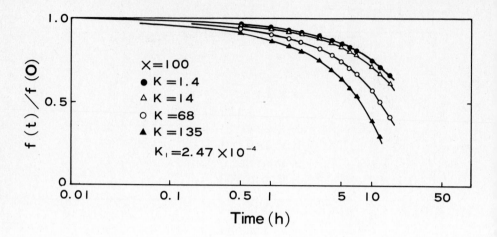

Fig. 3.6. Theoretical relationships between f(t)/f(O) and log t in eqn (3.2) as the function of k_1 and k_2 when x=100.

APPENDIX

We considered the case of the tetrahedral model as the smallest unit of perfect network structure. When scissions of the main chain and crosslinkage both occur at the same time in the usual crosslinked polymer having a perfect network structure, the theoretical treatment is as follows. As defined above, k(t) is the ratio of the probability of crosslink scission to that of main chain scission, consequently, k(t)=O when only main chain scission occurs, and k(t) = ∞ when only crosslinkage scission occurs.

The equations defined above are shown once again.

$$\varrho(t) = q_m(t) + q_c(t) \tag{3.9}$$

$$k(t) = \frac{q_c(t)/C(0)}{q_m(t)/M_o} \tag{3.3}$$

$$M_o = x \cdot n(0)$$

It is now assumed that the crosslinkage scission is unity or less than unity for the smallest unit of network structure as shown by Fig. 3.7.

The expression "scission ratio is less than unity" means that the number of crosslinkage scissions is less than that of the units of the smallest model in the plural network structure. The above assumption is expressed by eqn (3.10).

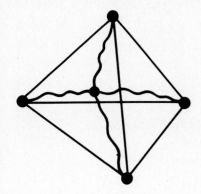

Fig. 3.7. A smallest unit model of perfect network structure.

$$\frac{q_c(t)}{n(0)/2} \leq \frac{1}{5} \qquad\qquad (3.10)$$

Substituting eqns (3.9) and (3.3) into eqn (3.10), eqn (3.10)a is obtained.

$$\frac{q_m(t)}{M_o} \leq \frac{1}{5} \left(\frac{1}{k(t)} + \frac{1}{2x}\right) \qquad\qquad (3.10)a$$

On the other hand, from eqns (3.9) and (3.3),

$$q_m(t) = \frac{2x}{2x+k(t)} \cdot Q(t) \qquad\qquad (3.11)$$

$$q_c(t) = \frac{k(t)}{2x+k(t)} \cdot Q(t) \qquad\qquad (3.12)$$

Though scissions of main chain and crosslinkage both occur at the same time, we assume for convenience of calculation that the scissions occur firstly at the crosslinkage, i.e., x-mer and 2x-mer network chains appear owing to the changes after t hours under the conditions of eqn (3.10). The number of x-mer chains is $n(0)-4q_c(t)=n_x(0)$ and that of 2x-mer chains is $2q_c(t)$. Next, considering the scission of the main chain, the number of x-mer chain scissions $q_x(t)$ is,

$$q_x(t) = q_m(t)\frac{x \cdot n_x(0)}{M_o}$$

(3.13)

and that of 2x-mer chain scissions $q_{2x}(t)$ is,

$$q_{2x}(t) = q_m(t)\frac{4x \cdot q_c(t)}{M_o}$$

(3.14)

Therefore, the number of residual x-mer chains after t hours, $n_x($ is obtained as follows.

$$n_x(t) = n_x(0)\left\{1 - \frac{q_x(t)}{n_x(0)}\right\}^x \quad n_x(0)e^{-q_x(t)/n_x(0)}$$

(3.15)

Similarly, the number of residual 2x-mer chains $n_{2x}(t)$ becomes,

$$n_{2x}(t) = 2q_c(t) \cdot \left\{1 - \frac{q_{2x}(t)}{4xq_c(t)}\right\}^{2x} \quad 2q_c(t) \cdot e^{-q_{2x}(t)/2q_c(t)}$$

(3.16)

The total number of effective network chains is

$$n(t) = n_x(t) + n_{2x}(t)$$

(3.17)

Consequently, the relative stress, $f(t)/f(0)$ can be expressed by eqn (3.18).

$$\frac{f(t)}{f(0)} = \frac{n(t)}{n(0)} = \frac{n_x(t) + n_{2x}(t)}{n(0)}$$

$$= \frac{1}{n(0)} \cdot n_x(0) \cdot e^{-q_x(t)/n_x(0)} + 2q_c(t)\ e^{-q_{2x}(t)/2q_c(t)}$$

(3.18)

Substituting eqns (3.11) and (3.16) into eqn (3.18),

$$\frac{n(t)}{n(0)} = \frac{f(t)}{f(0)}$$

$$= 1 - \frac{4k(t) \cdot x}{k(t)+2x} \cdot \frac{Q(t)}{M_o} \cdot \exp\left\{-\frac{2x}{k(t)+2x}\ \frac{Q(t)}{M_o}\right\}$$

$$+ \frac{2k(t) \cdot x}{k(t)+2x} \cdot \frac{\varrho(t)}{M_o} \cdot \exp\left\{ - \frac{2x}{k(t)+2x} \cdot \frac{\varrho(t)}{M_o} \right\} \tag{3.19}$$

That is, eqn (3.19) in the Appendix corresponds to eqn (3.2) in the above chapter. Substituting $k(t)=0$ into eqn (3.19), or in the case of main chain scission only, eqn (3.19)a is obtained.

$$\frac{n(t)}{n(0)} = \frac{f(t)}{f(0)} = e^{-q_m(t)/n(0)} \tag{3.19}a$$

On the other hand, substituting $k(t)=\infty$ in eqn (3.19), or in the case of crosslinkage scission only, eqn (3.19)b is derived.

$$n(0) = n(t) + 2\varrho(t) \tag{3.19}b$$

Eqns (3.19)a and (3.19)b are entirely in agreement with the equations proposed by A.V. Tobolsky (ref. 13).

Furthermore, with a similar model, but assuming that the crosslinkage scissions are more than unity but less than two under the conditions represented by eqn (3.20) below, eqn (3.21) is derived, which is somewhat more complicated than eqn (3.19).

$$\frac{1}{5} < \frac{q_c(t)}{n(0)/2} \leq \frac{2}{5} \tag{3.20}$$

$$\frac{n(t)}{n(0)} = \frac{f(t)}{f(0)} = \left\{ 1 - \frac{7}{2} \cdot \frac{k(t) \cdot x}{k(t)+2x} \cdot \frac{\varrho(t)}{M_o} \right\} \cdot \exp\left\{ - \frac{2x^2}{k(t)+2x} \cdot \frac{\varrho(t)}{M_o} \right\}$$

$$+ \frac{k(t) \cdot x}{k(t)+2x} \cdot \frac{\varrho(t)}{M_o} \cdot \exp\left\{ - \frac{4x^2}{k(t)+2x} \cdot \frac{\varrho(t)}{M_o} \right\}$$

$$+ \frac{k(t) \cdot x}{2k(t)+4x} \cdot \frac{\varrho(t)}{M_o} \cdot \exp\left\{ - \frac{6x^2}{k(t)+2x} \cdot \frac{\varrho(t)}{M_o} \right\} \tag{3.21}$$

Here, from the Table 3.1 in the above chapter, in our experiment for TMTD-cured natural rubber, $q_c(t)/n(0)/2$ is obviously very small and adequately satisfies the condition of eqn (3.10).

Consequently, it is clear that eqn (3.19) in the Appendix (or eqn (3.2) in the above chapter) is reasonable and satisfactory.

II EXCHANGE REACTION WITH CLEAVAGE AT CROSSLINKAGES AND/OR ALONG MAIN CHAINS

Ethylene-propylene terpolymer (EPT) was used as the sample materi in this experiment. The crosslinked polymer was prepared using cross linking agents and the methods of crosslinking specified in Table 3.

TABLE 3.2

Preparation of cured EPT polymers

	Sample 1	Sample 2	Sample 3
Rubber	100	100	100
Sulfur		2	
Zinc oxide		5	5
Stearic acid		2	1
Tetramethyl thiuram disulfide (TT)		1	3.5
Mercaptobenzothiazole (MBT)		0.5	
Dicumyl peroxide (DCP)	2		
Curing time, min*	30	20	40

* Curing temperature 150°C

Three kinds of EPT polymer, (samples 1, 2 and 3) differing only i the structure of the crosslinkages, were prepared. It is evident tha the crosslinking sites, in these samples, consist of carbon-carbon (C-C) bonds, a polysulfide linkages ($-S_x-$), and monosulfide linkages ($-S-$) for samples 1, 2 and 3, respectively.

The stress-relaxation of dicumyl peroxide-cured EPT (sample 1) wa measured at 109°C in air, in nitrogen, and in air containing a prope amount of antioxidant. These stress-relaxation curves are shown in

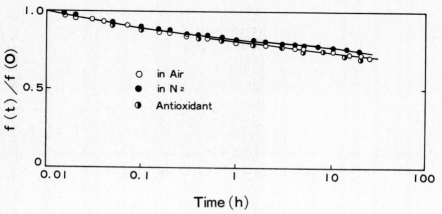

Fig. 3.8. Stress-relaxation of dicumyl peroxide-cured EPT in air, under nitrogen, and in air containing a known amount of antioxidant at 109°C.

Fig. 3.8. From this figure, we can see that the descending curves have almost the same slope in the three samples. When we consider both the chemical structures and the experimental conditions of sample 1, it seems very reasonable to assume that the relaxation mechanism of the sample is physical, that is, a physical flow of the polymer chains.

The stress-relaxation behavior of sample 1, with initial chain density $n(0)$ of 0.53×10^{-4} and 1.45×10^{-4} mol cm^{-3}, and samples 2 and 3, with $n(0)$ of 1.39×10^{-4} and 0.62×10^{-4} mol cm^{-3}, respectively, were also measured in air at 109°C. The relations between relative stress and time are shown in Fig. 3.9. Because Figs. 3.8 and 3.9 were obtained

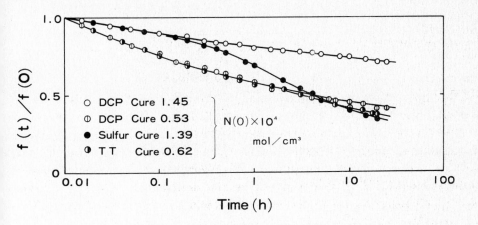

Fig. 3.9. Stress-relaxation of samples 1, 2 (sulfur-cured EPT), and 3 (TT-cured EPT) in air at 109°C.

under the same conditions, the portions of the curves for samples 2 and 3 which superpose with that of sample 1 are considered to be due to physical relaxation. From Fig. 3.9 it was deduced that the stress-decay due to a physical relaxation at the initial stages is independent of the kind of crosslink, if the crosslinking density of the polymers is equivalent. For further identification, we have carried out experiments with materials with different kinds of crosslinks and with identical initial densities $n(0)$. In these materials, the cross-links are carbon-carbon (C-C) bonds and polysulfide linkages ($-S_x-$), and the $n(0)$ values are 0.78×10^{-4}, 1.39×10^{-4}, and 2.26×10^{-4} mol cm^{-3}. The stress-relaxation curves obtained from these samples are shown in Fig. 3.10. From Fig. 3.10, it is clear that with nearly equal values of $n(0)$ in these samples, the physical relaxations in the initial stages are the same and overlap, regardless of the kind of

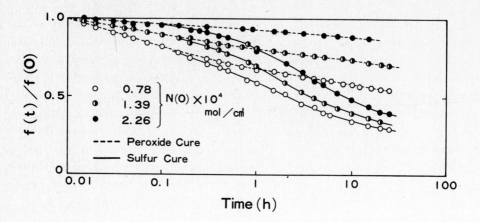

Fig. 3.10. f(t)/f(O) vs. time of three samples having different values of n(O) in air at 109°C.

crosslink.

In order to discuss the mechanisms of the chemical relaxation in detail, the portions of the stress-decay based upon physical flow were removed from the original stress-relaxation curves of EPT polymers; for samples cured with TMTD (samle 3, Fig. 3.9) and with sulfur (samp 2, Fig. 3.9 and 3.10), their pure chemical stress relaxation curves were replotted (Fig. 3.11 and 3.12) for reference.

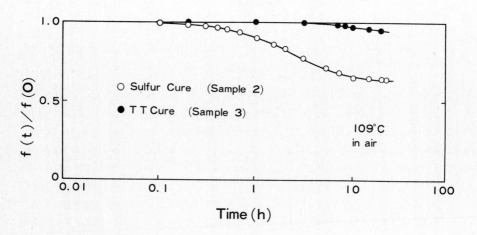

Fig. 3.11. Pure chemical stress-relaxation of samples 2 and 3 obtained from Fig. 3.8.

It can be easily assumed from Fig. 3.11 and 3.12 that few cleavage occur at the crosslink in sample 3, and the interchange reactions

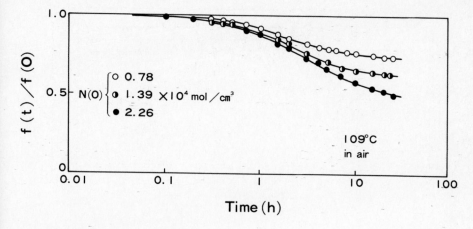

Fig. 3.12. Pure chemical stress-relaxation of sample 2 obtained from Fig. 3.10.

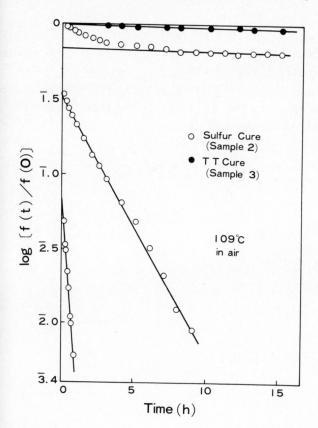

Fig. 3.13. Replot as log [f(t)/f(O)] vs. time of two chemical stress-relaxation curves in Fig. 3.10.

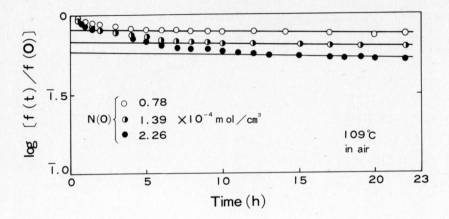

Fig. 3.14. log [f(t)/f(0)] vs. time of the sample 2 having a different value of n(0) in air at 109°C.

mostly occur at the crosslinking sites in sample 2.

　　Then, when the two chemical stress-relaxation curves in Fig. 3.11 were replotted as log [f(t)/f(0)] versus time, Fig. 3.13 was obtained. From such a plot, a fine straight line was obtained for sample 3. Because there still remained a curved portion of the plot in the short-time region for sample 2, we used our suggested procedure (ref. 2) and repeated it until we obtained a straight line over the whole

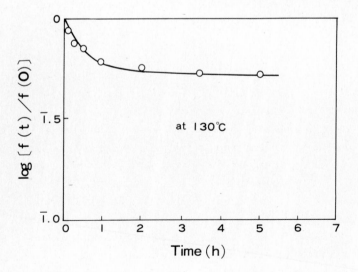

Fig. 3.15. Tobolsky et al. data for stress-relaxation of sample 2 in air at 130°C.

time-scale. When the three curves of Fig. 3.12 were replotted as log
[f(t)/f(O)] versus time, as in Fig. 3.13, three straight lines having
the same slope in the long-time region were obtained, as shown in
Fig. 3.14.

A.V. Tobolsky et al. (ref. 3), recently carried out chemical
relaxation measurements on sample 2, under similar experimental conditions to ours; the experimental results they obtained are shown in
Fig. 3.15. This figure indicates that the stress-decay based on the
interchange mechanism of a polysulfide crosslinkage occurs rapidly
at the initial stage of the relaxation time; after some hours, however,
the stress-decay curve becomes parallel to the abscissa because of
thermally-stable monosulfide and disulfide crosslinkages. These results
differ from ours in Fig. 3.13 and 3.14, and these differences merit
comment.

I) In our experimental results, that is, in the case of sample 2 in
Fig. 3.13 and 3.14, the straight lines for long times are not parallel
to the abscissa but have slightly descendent similar slopes. Which
results are correct?

II) The stress-relaxation curve relating to sample 2 is divided
into three straight lines by using Procedure X as shown in Fig. 3.13.
Can this be explained?

III) In Fig. 3.13, the slopes of the straight line for sample 3
and of the uppermost straight line of the three for sample 2 are
indentical. In Fig. 3.14, all the slopes of the three straight lines
for sample 2 are identical, independent of the crosslinking density
n(O). Why?

These questions may be resolved by the following considerations:

When scission occurs only at the crosslink in crosslinked polymers,
it is obvious from eqn (2.113) that the stress-relaxation curve is
expressed by a Maxwellian decay. On the other hand, it has been
ascertained from section of 2.3. that the stress-relaxation mechanism
of crosslinked polymers which undergo an interchange reaction can be
expressed by the sum of two exponential terms. Now, in the case of
the sulfur-cured EPT polymer (sample 2), it is assumed from the various
experimental results described that the crosslinking sites of this
sample consist of polysulfide, monosulfide, and disulfide linkages.
We can deduce for sample 2 that two reactions occur simultaneously at
the crosslinkage: typical interchange and cleavage. Therefore, the
relationship between relative stress and time can be expressed (ref.
4).

108

$$f(t)/f(0) = A \cdot e^{-k_1 t} + B \cdot e^{-k_2 t} + C \cdot e^{-k_3 t} \qquad (3.22)$$

The first two terms on the right-hand side of eqn (3.22) are due to the mechanisms of typical interchange reactions, and the third term refers to cleavage at the crosslinkage.

The question of why all the slopes of the first straight line in the sample 2 and the ones in the sample 3 are equal and independent of n(0) may now be answered on the basis of eqn (2.113) or the third term in eqn (3.22). According to Tobolsky et al. (ref. 5), in the case when scission occurs only at crosslinks in the crosslinked polymer, it is recognized that the decay curves of f(t)/f(0) versus time are identical and independent of the chain density n(0). Therefore the value of k in eqn (2.113), or of the third term in eqn (3.22) is usually a universal constant and is independent of n(0). As for the question of why there are three straight lines for sample 2, it is easy to understand that for one straight line there is one term in equation (3.22). Finally, it is very clear that the straight lines in the long-time region are not parallel to the abscissa because of the slope k of the straight line.

The results of Fig. 3.16 were obtained by us (ref. 6) for the four

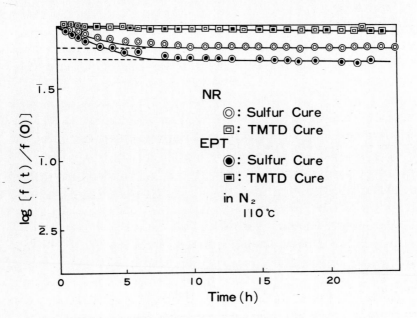

Fig. 3.16. The relation between log [f(t)/f(0)] vs. time for natural rubber vulcanizates with sulfur and TMTD, and EPT vulcanizates with sulfur and TMTD in N_2 at 110

samples in nitrogen at 110°C. The four samples were natural rubber
and EPT cured with sulfur or with TMTD. It was found that the straight
line for TMTD-cured natural rubber is not only quite consistent with
that for TMTD-cured EPT, as shown in Fig. 3.15, but also with the
straight line for TMTD-cured EPT in air at 110°C (Fig. 3.12). These
results indicate that scission occurs at crosslinks in the two samples,
as well as in TMTD-cured EPT in air at 110°C. The result for TMTD-
cured natural rubber in N_2 is consistent with that obtained in vacuo by
Takahashi et al. (ref. 7). Furthermore, both slopes of the straight
lines in the long-time region for sulfur-cured natural rubber and
that for sulfur-cured EPT are identical with those for TMTD-cured
samples, as indicated in Fig. 3.15. From the above results, it is
clear that scission at the crosslinks occurs for sulfur-cured natural
rubber and EPT in N_2 together with the exchange reaction of polysulfide
linkages. These two samples are respectively divided into three
straight lines by procedure X; the uppermost one indicates scission
of crosslinks and the other two show exchange scission.

The mechanisms of scission at crosslinkages and exchange scission
were dealt with above for natural rubber vulcanizates in N_2 and EPT
vulcanizates in air and N_2. The mechanism of scission for natural
rubber vulcanizates cured by different crosslinking methods in air
will be discussed as follows.

Accelerated sulfur-cured natural rubber (sample 6) and irradiation-
sulfur-cured natural rubber (samples 7 and 8) were prepared under
conditions described in Table 3.3. Other conditions of preparation
were same as for the samples in Table 2.4 in section of 2.1.5.

TABLE 3.3

Preparation of cured NR polymers

	Sample		
	6	7	8
Rubber	100	100	100
Sulfur	3	3	3
Zinc oxide	3	3	–
Stearic acid	0.5	0.5	–
TMTD	–	–	–
Mercapto benzothiozole (MBT)	1	1	–
DCP	–	–	–
Hot-press curing*	30 min.	30 min.	–
Irradiation curing+	–	+	+

* Curing temperature 145°C
+ Total dose=12, 28.8 and 43.2 Mrad

110

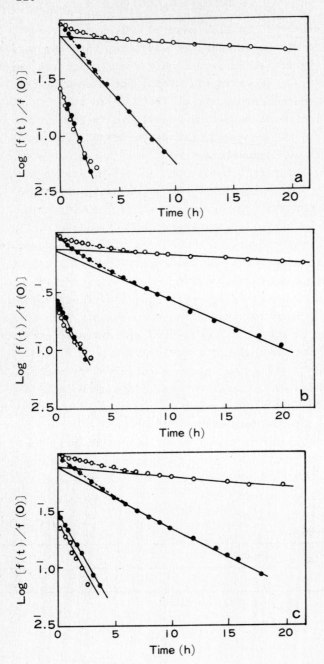

Fig. 3.17. (a), (b), (c) Stress-relaxation in air and under nitrogen at 100°C. ● , in air; ○ , under nitrogen, (a) sample 6; (b) sample 7 prepared by irradiation with 28.8 Mrad; (c) sample 8 prepared by irradiation with 28.8 Mrad.

Stress-relaxation of samples 7 and 8 was measured in both air and N_2 at 100°C. The results are shown in Fig. 3.17(a) for sample 6, Fig. 3.17(b) for sample 7 and Fig. 3.17(c) for sample 8 (ref. 8). Samples 7 and 8 were prepared using an irradiation dose of 28.8 Mrad. Cross-linkages in these samples consist of C-C bonds and mono-, di- and polysulfide linkages, as described previously. From Fig. 3.17, it can be seen that all the stress-relaxation curves decay rapidly in the short-time region and then become linear at longer times. We used Procedure X and repeated it until we obtained a straight line over the whole time-scale. Either in air or under nitrogen, all the stress-relaxation curves for samples 6, 7 and 8 were divided into two definite straight lines, as shown in Fig. 3.17.

It is evident from the figures that the stress-relaxations in the short-time region for all samples are in fair agreement with each other in both air and N_2, whereas the one in the long-time region is apparently different as between air and N_2. In more detail, the slope of the upper straight lines (shown by white marks) in N_2 in Fig. 3.17 was seen to be constant, independent of the crosslinking density n(0) and crosslinking structure. As for the lowest straight lines, both the ones in air and N_2 in each figure are fairly consistent with each other. The middle straight lines (shown by black marks) in air are found to change, depending on n(0). These results indicate that the upper straight lines in N_2 denote scission at crosslink sites due to the cleavages of mono and/or disulfide linkages in natural rubber vulcanizates, the lowest ones (in N_2 and air) denote exchange scission, and the middle ones (in air) indicate scission along main chains. Thus, the above can be expressed by:

$$\frac{f(t)}{f(0)} = A\exp(-k_A t) + B\exp(-k_B t) \tag{3.23}$$

$$(\text{in air})$$

$$\frac{f(t)}{f(0)} = A'\exp(-k_A t) + B'\exp(-k_B' t) \tag{3.24}$$

$$(\text{in nitrogen})$$

where,

$A\exp(-k_A t)$, $A'\exp(-k_A t)$: exchange reactions,

$B\exp(-k_B t)$: scission along main chains, $\tag{3.25}$

112

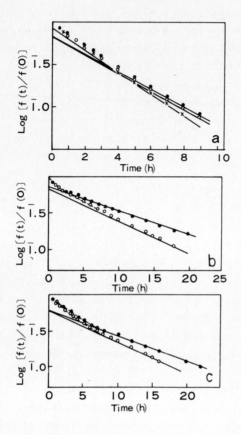

Fig. 3.18. (a), (b), (c) Dependence on initial chain-density n(0) of the stress-relaxation in air at 100°C for samples 6, 7 and 8. (a) No. 6: x, 1.10×10^{-4}; o, 1.31×10^{-4}; ●, 1.46×10^{-4} mol cm^{-3}; (b) No. 7: o, 2.45×10^{-4}; ●, 2.84×10^{-4} mol cm$^-$ (c) No. 8: o, 0.85×10^{-4}; ●, 1.13×10^{-4} mol cm^{-3}.

Fig. 3.18(a), 3.18(b), and 3.18(c), respectively, show the dependence on initial chain density, n(0) of the stress-relaxation in air for samples 6, 7 and 8. Those in nitrogen are shown in Fig. 3.19(a) and 3.19(b) for the purpose of comparison with samples 3, 4 and 5 of Table 2.4 in section of 2.1.5. From the figures, it can be seen that the k_B values in air depend on n(0), while the k_B' value is independer of n(0).

From eqns (3.23) and (3.24), it is expected that there will be two different stress-relaxation mechanisms in air and in nitrogen;

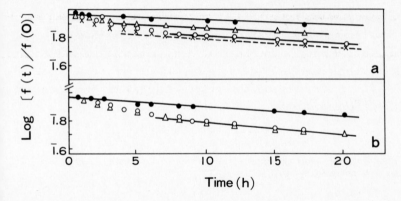

Fig. 3.19. (a), (b) Stress-relaxation of samples 3, 4, 5, 6, 7 and 8 under nitrogen at 100°C. (a) ●, 3 and 4; ×, 6; △ and ○, 7. (b) ●, 5; △ and ○, 8.

one is independent of the surroundings, the other will have completely different mechanisms between air and nitrogen. These stress-relaxation behaviors can be explained as follows. The stress-relaxation (k_A) in the short-time region will be based on the interchange reactions of polysulfide linkages which are independent of the surroundings. The exchange reactions of polysulfide linkages have been investigated in detail by Tobolsky (ref. 9), Beevers (ref. 10), and ourselves (ref. 10), as described in the section of 2.3. According to them, the exchange reactions should be expressed by the sum of two Maxwellian decay terms. From the experimental point of view, however, one term may usually be enough. Previously, we reported (ref. 12) that the stress-relaxation in nitrogen of accelerated-sulfur-cured natural rubber (sample 6) was based on the exchange reactions of polysulfide linkages and the thermal scission of mono- and disulfide linkages. All the slopes (k_B') of straight lines for sample 7 in the long-time region are not only independent of the initial chain density n(0), but are also equal to that of TMTD-cured natural rubber in which the crosslinks consist of mono- and disulfide linkages. The slopes for sample 8 are also independent of n(0) and are in agreement with those for sample 5.

It can thus be concluded that the stress-relaxation in nitrogen of samples 7 and 8 in the long-time region is also based on the thermal scission of crosslinkages. On the other hand, stress-relaxation (k_B) in air of samples 6, 7 and 8, apparently depends on the initial chain density n(0). Therefore, these vulcanizates also appear to undergo oxidative scission of the main chains like the others. In the cured systems, the presence of mono-, di-, and polysulfide linkages also appears to accelerate oxidative scission of the main chain.

In conclusion, oxidative and thermal degradation mechanisms for natural rubber vulcanizates which have carbon-carbon bonds, carbon-monosulfur-carbon bonds, and carbon-polysulfur-carbon bonds as cross links are classified and tabulated in Table 3.4.

TABLE 3.4

Degradation mechanisms at high temperature T for natural rubber vulcanizates with different crosslinking density n(0)

Kind of Polymer / Degradation	Structure	Thermal Degradation	Oxidative Degradation
Peroxide or Irradiation cured Natural Rubber	$CH_3 \quad CH_3$ $-C-CH_2-CH_2-C-CH-CH_2-$ $-C-CH_2-CH_2-C-CH-CH_2-$ $CH_3 \qquad CH_3$	Random Scission along Main Chains. $\dfrac{f(t)}{f(0)}=e^{-\dfrac{q(t)}{n(0)}}$ $(q(t)\langle q'(t))$	Random Scission along Main Chains $\dfrac{f(t)}{f(0)}=e^{-\dfrac{q'(t)}{n(0)}}\approx e^{-k_1 t}$ $k_1=f\ \{n(0)\}\ (q(t)\langle q'(t))$
TMTD cured Natural Rubber	$CH_3 \quad CH_3$ $-C-CH_2-CH_2-C=CH-CH_2-$ S $-C-CH_2-CH_2-C=CH-CH_2-$ $CH_3 \qquad CH_3$	Scission at Crosslinks. $\dfrac{f(t)}{f(0)}=e^{-k_2 t}$ $k_2=$const.	Random Scission along Main Chains. $\dfrac{f(t)}{f(0)}=e^{-k'_1 t}$ $k'_1 \rangle k_1$ $k'_1=g\ \{n(0)\}$
Sulfur cured Natural Rubber	$CH_3 \quad CH_3$ $-C-CH_2-CH_2-C=CH-CH_2-$ Sx $-C-CH_2-CH_2-C=CH-CH_2-$ $CH_3 \qquad CH_3$	Scission at Crosslinks and Exchange Reaction. $\dfrac{f(t)}{f(0)}=A e^{-k_2 t}+B e^{-k_3 t}$ $k_2=$const.	Random Scission along Main Chains and Exchange Reaction. $\dfrac{f(t)}{f(0)}=A e^{-k_2 t}+B e^{-k'_1 t}$

REFERENCES

1 K. Murakami, S. Tamura, Y. Takano and H. Kurumiya, Kobunshi Kagaku (Chem. High Polym., Japan), 24(1967)699; K. Murakami and S. Tamura, Polym. J., 2(1971)328; K. Murakami and S. Tamura, J. Polym. Sci., 9(1971)423.
2 A.V. Tobolsky and K. Murakami, J. Polym. Sci., 40(1959)443; A.V. Tobolsky, Properties and Structure of Polymers, Wiley, New York, 1950, p.188.
3 A.V. Tobolsky, J. Polym. Sci., B, 2(1964)823; R.B. Beevers, J. Colloid Sci., 19 (1964)40.
4 K. Murakami and S. Tamura, J. Polym. Sci., 9(1971)423; K. Murakami and S. Tamura, Polym. J., 2(1971)328.
5 A.V. Tobolsky, J. Appl. Phys., 27(1956)673.
6 K. Murakami and S. Tamura, J. Polym. Sci., B. 11(1973)529.
7 Y. Takahashi and A.V. Tobolsky, Polym. J., 2(1971)457.
8 S. Tamura and K. Murakami, Polym. 14(1973)569.
9 A.V. Tobolsky, J. Polym. Sci., B, 2(1964)823.
10 R.B. Beevers, J. Colloid Sci., 19(1964)40.
11 K. Murakami and S. Tamura, J. Polym. Sci., A-1, 9(1971)423.
12 K. Murakami and S. Tamura, J. Polym. Sci., B, 11(1973)317.
13 A.V. Tobolsky, Properties and Structure of Polymers, Wiley, New York, 1960, Chapter V.

116

CHAPTER 4

CROSSLINKING REACTION

I MECHANISM OF CROSSLINKING REACTION

1 Standard Method of Intermittent Measurement

It was observed that, during the chemical stress-relaxation experi ments in the temperature range 100 to 150°C, some of the rubber vulcanizates, such as Butyl, became progressively softer while others such as GR-S, showed a continued hardening, and even others, such as natural rubber, first softened and then hardened. It was deduced that, at these elevated temperatures, crosslinking and scission were occurring simultaneously, and the rates of both crosslinking and scission were essentially unaffected by the stress or elongation of the sample.

We shall assume that a sample of rubber vulcanizate is left at a given high temperature in air for oxidative degradation. The sample at the initial stage (t=0) is shown in Fig.4.1A,i. Suppose, for simplicity, that no scission reaction occurs but that crosslinking occurs to some extent after time $t=t_1$, as shown in Fig. 4.1A,ii.

At widely spaced time intervals, $t=t_1$, the rubber is rapidly stretched to a fixed elongation (say a%), the equilibrium stress, f_1, is rapidly measured during the minimum time, $t=t_2$, as shown in Fig. 4.1A,iii and iv and the sample is immediately returned to its unstretched length, as indicated in Fig. 4.1A,v. Though some addi- tional crosslinkages, denoted by the dotted lines in the figure, are seen to be formed, even during the minimum time $t=t_2$, such new cross- linkages do not, to a first approximation, affect the stress of the sample held at constant length, unless the crosslinking becomes so extensive that there is volume shrinkage.

Again, at widely spaced time intervals, $t=t_3$, the method described above is repeated as seen in Fig. 4.1A,vi, vii, viii, and iv, and so on.

Nearly all the scission and crosslinking may be considered as occurring while the network is in the unstretched condition. The intermittent stress measurements reflect the combined effect of scission and crosslinking. Therefore, it is possible to measure the

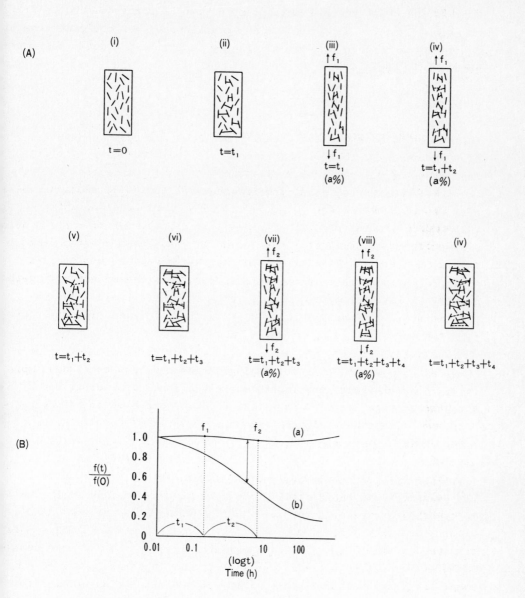

Fig. 4.1.Mechanisms of continuous and intermittent stress-relaxation measurement.
(A) Intermittent stress-relaxation measurement. (B) Continuous and intermittent
stress-relaxation curves. (a) Intermittent (b) Continuous.

total number of additional crosslinkings and chain scissions separately
by using curves (a) and (b) of Fig. 4.1B which are derived in the
unstretched condition.

Suppose that the original network structure is characterized by C(O) moles of effective tetrafunctional crosslinkages per cubic centimeter The relation between n(O) moles of network chains per cubic centimete and C(O) is given by,

$$n(0) = 2C(0) \tag{2.112}$$

The number of moles of additional crosslinks per cubic centimeter tha have subsequently formed up to time t will be denoted by $\varDelta C(t)$, and the number of moles of chain cleavages per cubic centimeter by $q(t)$. Then the crosslinking density of the network chains up to time t, $n(t)$ is given by,

$$n(t) = 2C(t) = 2C(0) - q(t) + 2 \varDelta C(t) \tag{4.1}$$

Therefore the intermittent stress measurement at time t, $f_i(t)$ is denoted by,

$$f_i(t) = \{2C(0) - q(t) + 2 \varDelta C(t)\} \, RT(\alpha - \alpha^{-2}) \tag{4.2}$$

On the other hand, the stress $f_c(t)$ measured in a continuous relaxation of stress experiment, where the sample is maintained at a fixed elongation, is

$$f_c(t) = \{2C(0) - q(t)\} \, RT(\alpha - \alpha^{-2}) \tag{4.3}$$

If the difference between curves (a) and (b) in Fig. 4.1B is represented by X, the following equation is derived from eqns (2.112) (4.1), (2.2), (4.2) and (4.3).

$$X = \frac{f_i(t)}{f(0)} - \frac{f_c(t)}{f(0)} = \frac{2 \varDelta C(t)}{n(0)} = \frac{\varDelta C(t)}{C(0)} \tag{4.4}$$

From the difference value X in Fig. 4.1B, the additional cross- linkages formed up to time t was found from eqn (4.4). If only scissi were occurring, the continuous stress-relaxation curves and the intermittent stress curves would be identical. The above method can be applied for various kinds of rubber vulcanizates. (refs. 1,2)

2 Improved Method of Intermittent Measurement by Ore (ref. 3) (ITMM Method)

Ore proposed an improved continuous stress-relaxation method for the purpose of measuring the amount of crosslinkages in the elongated state and at low elongation, as follows.

If the stress in the continuous method is denoted by $f_c(t)$, and the crosslinking density by $n_o(t)$ after time t, eqn (2.2) is rewritten as

$$f_c(t) = n_o(t)RT(\alpha - \alpha^{-2}) \qquad (4.5)$$

If the extension ratio of a sample is increased by a small amount $d\alpha$, the increased stress $df_c(t)$ will be denoted by differentiation of eqn (4.5) with respect to α, as follows.

$$df_c(t) = RT(1 + \frac{2}{\alpha^3})n_o(t) \cdot d\alpha \qquad (4.6)$$

On the other hand, the increase of extension ratio, $d\alpha$ leads to an increase in the stress due to $n_\alpha(t)$, the moles of additional new network chains per cubic centimeter which are formed up to time t. As the sample having such new chains $n_\alpha(t)$ is in the unstretched state at elongation α, the increased stress $df_\alpha(t)$ due to $n_\alpha(t)$ will be given by eqn (4.7) for a sample with cross-sectional area A_o/α.

$$df_\alpha(t) = \frac{1}{\alpha} \cdot RT \{1 + \frac{d\alpha}{\alpha} - \frac{1}{(1 + \frac{d\alpha}{\alpha})^2}\} \cdot n_\alpha(t)$$
$$= RT(\frac{3}{\alpha^2}) \cdot n_\alpha(t) \cdot d\alpha \qquad (4.7)$$

Therefore, from eqns (4.6) and (4.7), the total increase of stress $df_i(t)$, equal to $\{df_c(t) + df_\alpha(t)\}$, is given by

$$df_i(t) = RT \{(1 + \frac{2}{\alpha^3})n_o(t) + \frac{3}{\alpha^3} \cdot n_\alpha(t)\} d\alpha \qquad (4.8)$$

The increasing ratio $df_i(t)/d\alpha$, due to the increase of extension ratio, assuming that α takes the values 1.20 and 1.50, is given by,

$$(\frac{df_i(t)}{d\alpha})_{\alpha=1.20} = RT \{2.16n_o(t) + 2.08n_\alpha(t)\} \qquad (4.9)$$

$$(\frac{df_i(t)}{d\alpha})_{\alpha=1.50} = RT \{1.59n_o(t) + 1.33n_\alpha(t)\} \qquad (4.10)$$

The stress calculated by the intermittent relaxation method $f_i(t)$ at $\alpha=1.20$ and $\alpha=1.50$ is given by the following equations, while the stress obtained by Tobolsky's intermittent method is given by eqn (4.2),

$$\{f_i(t)\} = RT \{0.51n_o(t) + 0.51n_\alpha(t)\} \qquad \qquad (4.11)$$
$$\alpha=1.20$$

$$\{f_i(t)\} = RT \{1.06n_o(t) + 1.06n_\alpha(t)\} \qquad \qquad (4.12)$$
$$\alpha=1.50$$

where $n_o(t)$, in the theory of Ore, corresponds to $2C(O)-q(t)$ in the theory of Tobolsky, and $n_\alpha(t)$ to $2\Delta C(t)$, similarly.

From inspection of eqns (4.9), (4.10), (4.11) and (4.12), it was found that the two quantities representing initial network chains, $n_o(t)$ and additional network chains, $n_\alpha(t)$ contribute to $f_i(t)$ equal but not to $df_i(t)/d\alpha$. The contribution of $n_\alpha(t)$ to $df_i(t)/d_\alpha$ is abou 4% different from that of $n_o(t)$, when the extension ratio is compara tively small, i.e. $\alpha=1.20$. The difference, however, becomes approxi mately 20% at $\alpha=1.50$.

The intermittent measurement of stress relaxation by Ore will be carried out as follows. The small stress ΔF must often be added to a sample by a small loading to the balance of a relaxometer, while the continuous stress-relaxation measurement is carried out. This result in an increase in the extension ratio of the sample $\Delta\alpha$, which can be measured by a microscope. Such added stress ΔF is adjusted so that $\frac{\Delta\alpha}{\alpha}\approx\frac{1}{10}$, according to Ore's experiments.

The chemical stress-relaxation curves by continuous and intermitt methods for natural rubber vulcanizates purified with acetone at 110 in air at $\alpha=1.20$ are shown in Fig. 4.2. From the figure, the normal intermittent stress $f_i(t)/f_i(O)$ is seen to be consistent with the intermittent tangent modulus $(\Delta F/\Delta\alpha)_{t=t}/(\Delta F/\Delta\alpha)_{t=O}$ for all ranges of the experiment. The chemical stress-relaxation curves for unextracte natural rubber vulcanizates at 110°C in air at $\alpha=1.50$ are shown in Fig. 4.3, which indicates no difference between $f_i(t)/f_i(O)$ and $(\Delta F/\Delta\alpha)_{t=t}/(\Delta F/\Delta\alpha)_{t=O}$. The ratio of initial stress to intermittent stres is nearly unity, independent of time, showing that the rate of chain scission is approximately equal to that of additional chain-forming by crosslinking reactions, for the reason discussed in section 2.1.1

In conclusion, the intermittent tangent modulus measurement metho (ITMM method), proposed by Ore, will be suitable for di-t-butyl

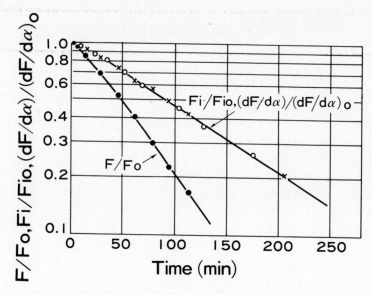

Fig. 4.2.Continuous and intermittent stress-relaxation curves for natural rubber vulcanizates (purified) at 110°C ($\alpha=1.20$). O Intermittent (Ore's method)
x Intermittent (Tobolsky's method) ● Continuous.

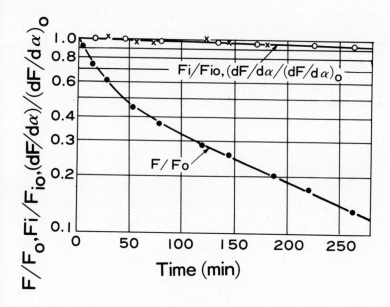

Fig. 4.3.Continuous and intermittent stress-relaxation curves for natural rubber vulcanizates (not extracted) at 110°C ($\alpha=1.50$). O Intermittent (Ore's method)
x Intermittent (Tobolsky's method) ● Continuous.

hydroperoxyde (DTBP)-cured natural rubber, because this rubber is

122

not good for heat-resistance; the normal intermittent method is dif
ficult to apply as the sample is liable to break after only a few
percent elongation.

3 Simultaneous Measurement for Continuous and Intermittent Relaxati
(SMCIR Method)

The ITMM method of Ore was found difficult to apply to stress-
relaxation apparatus of the strain-gauge type; measurements with
small loads and small extension ratios are also found difficult fro
the experimental point of view. We now describe (ref. 4) how to der
intermittent stress-relaxation curves by double elongation of a sam
as in Ore's method.

When a sample is stretched from its original length L_o to an
elongated length L_1 and maintained in this state, the relaxation st
after time t is given by eqn (4.13) which is derived from eqn (4.3)

$$f_c(t) = \{2C(0) - q(t)\}\ RT(\alpha_1 - \alpha_1^{-2}) \tag{4.13}$$

Here C(0) is the initial crosslinking density, q(t) the moles of ch
scission after time t, and α_1 the extension ratio expressed by $\alpha_1 = L$

Furthermore, the sample is stretched from L_1 to L_2 and the stre
$f_i(t)$ is measured in the latter state. The sample, in which the net
chains number $\{2C(0)-q(t)\}$ after time t, is now at an extension rat
$\alpha_2 = L_2/L_o$, but the additional network chains formed after time t i
the same sample [and which number $2 \Delta C(t)$] are regarded as being at
an extension ratio $\alpha_2/\alpha_1 = L_2/L_1$ because they exist subsequent to
attainment of the equilibrium state corresponding to L_1. $f_i(t)$ is
given by,

$$f_i(t) = \{2C(0)-q(t)\}\ RT(\alpha_2-\alpha_2^{-2}) + 2 \Delta C(t)\ RT\{\frac{\alpha_2}{\alpha_1} - (\frac{\alpha_1}{\alpha_2})^2\} \tag{4.14}$$

The stress at α_1 and α_2 for initial crosslinking density C(0) is
given by

$$f_c(0) = 2C(0)\ RT(\alpha_1-\alpha_1^{-2}) \tag{4.15}$$

$$f_i(0) = 2C(0)\ RT(\alpha_2-\alpha_2^{-2}) \quad \text{for } \alpha_1,\ \alpha_2 \text{ respectively.} \tag{4.16}$$

From eqns (4.15) and (4.16),

$$f_i(0) = \frac{\alpha_2 - \alpha_2^{-2}}{\alpha_1 - \alpha_1^{-2}} \cdot f_c(0) \tag{4.17}$$

The difference between the intermittent stress-relaxation curve
and the continuous curve is given by $\{f_i(t)/f_i(0)\} - \{f_c(t)/f_c(0)\}$ and one
obtains using eqns (4.13), (4.14), (4.15), (4.16) and (4.17):

$$\frac{f_i(t)}{f_i(0)} - \frac{f_c(t)}{f_c(0)} = \frac{\Delta C(t)}{C(0)} \frac{\frac{\alpha_2}{\alpha_1} - (\frac{\alpha_2}{\alpha_1})^2}{\alpha_2 - \alpha_2^{-2}} \tag{4.18}$$

The right-hand side of eqn (4.18) becomes $\Delta C(t)/C(0)$ when $\alpha_1 = 1$,
and conforms to the equation of Tobolsky's standard method, as
described in section 4.1.1.

When the above theory is applied, the continuous stress-relaxation
curve corresponding to α_1 and the intermittent curve corresponding
to α_2 are obtained at the same time for one sample, and the amount
of crosslinkages formed can be estimated from eqn (4.18).

The procedure by which continuous and intermittent relaxation
curves are obtained at the same time for one sample is called simultaneous measurement for continuous and intermittent relaxation (the
SMCIR method). Details of the SMCIR method are as follows.

The stress-relaxation after the sample is stretched to L_1 from
the original length L_o is measured and after time t_1 the sample is
stretched to L_2 from L_1 to measure the next phase of stress-relaxation.
After the equilibrium stress at L_2 has been rapidly measured, the
sample is immediately returned to L_1, and measurement of the continuous
stress-relaxation is resumed at L_1. Thus the continuous and intermittent measurements can be repeated in this fashion.

The SMCIR method has the following advantages over Tobolsky's
method.

(1) The SMCIR method can minimize the errors caused in the measurement compared to Tobolsky's normal method which needs two samples and
two conditions in the measurement; one sample and one condition are
enough in the present method.

(2) The measurement of crosslinkages formed in the stretched states
is possible in this method and information about the effect of crosslinking reactions on the stress measured is obtained.

(3) The assumption involved in Tobolsky's method that the total

number of network chain-scissions is the same in the unstretched an
stretched states is unnecessary for this method, because the amount
of network chains formed by crosslinking is estimated at L_1.

Our experimental results for the continuous and intermittent cur
for natural rubber and EPT vulcanizates obtained by SMCIR method ar
shown in Figs. 4.4, 4.5 and 4.6.

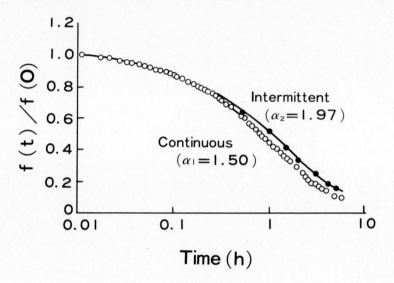

Fig. 4.4.Continuous and intermittent stress-relaxation curves for natural rubber
vulcanizates obtained by the SMCIR method.

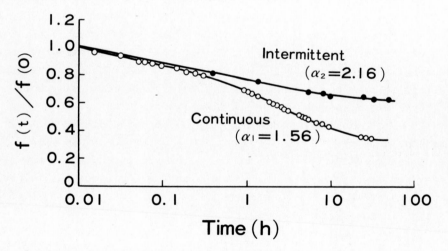

Fig. 4.5.Continuous and intermittent stress-relaxation curves for EPT rubber
vulcanizates obtained by the SMCIR method ($\alpha_1=1.56$, $\alpha_2=2.16$).

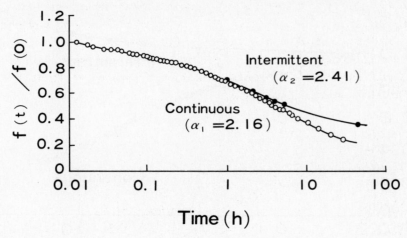

Fig. 4.6.Continuous and intermittent stress-relaxation curves for EPT rubber vulcanizates; SMCIR method $(\alpha_1=2.16, \alpha_2=2.41)$.

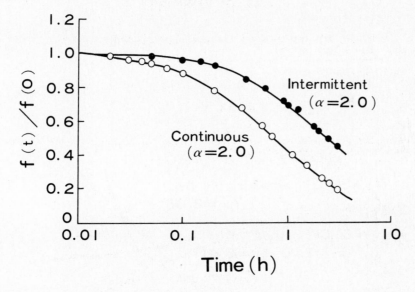

Fig. 4.7.Continuous and intermittent stress-relaxation curves for natural rubber vulcanizates; Tobolsky's method.

Similarly, the continuous and intermittent curves for the same samples derived using the Tobolsky method are shown in Figs. 4.7 and 4.8.

That the differences between two curves in Figs 4.4, 4.5 and 4.6 are smaller than those in Figs. 4.7 and 4.8 is considered to be related to the correction term linking eqns (4.4) and (4.18).

The relative number of additional crosslinkages formed, $\Delta C(t)/C(0)$,

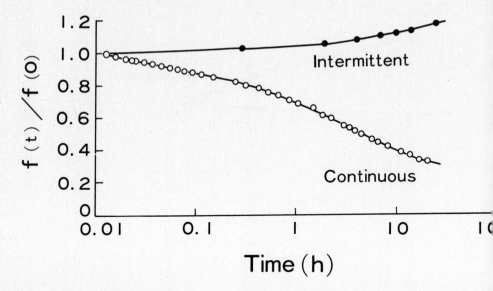

Fig. 4.8. Continuous and intermittent stress-relaxation curves for EPT rubber vulcanizates; Tobolsky's method.

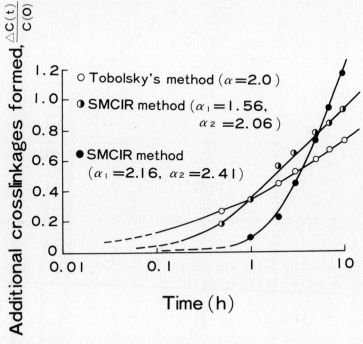

Fig. 4.9. The effect of extension on the amount of additional crosslinkages formed after time t. $\Delta C(t)$: Additional crosslinkages formed after time t, $C(O)$: Initial crosslinkages.

is plotted against time t for EPT vulcanizates in Fig. 4.9.

II CROSSLINKING REACTIONS AT LOW TEMPERATURES

Polymers can change their molecular structures under mechanical stimulus, and these changes result from cleavage of the polymer chains. Generally, the average molecular weights of masticated elastomers are smaller than those of the unmasticated specimens. If other reactive components are present, the unstable free radicals will combine with these substances, in so-called mechano-polymerization. The portions of polymer chains changed by mechanical stimulus are chemically unstable and always react so as to attain stable structures. Some studies on mechano-polymerization have recently been reported (refs. 5-7).

It is well-known that when polymers are irradiated, gaseous hydrogen is liberated from the points of rupture of the broken molecular chains and crosslinking reactions are induced subsequently. Similar phenomena have been found in the case of mechanical molecular cleavage, and these mechanisms can be illustrated by the following reaction (ref. 8).

$$(1)$$

$$(2)$$

When vulcanized rubbers (having almost zero mobility and a large coefficient of friction between molecular chains), are forced to extend in the glassy state or near it, scissions will be considerable. The recombination of ruptured main chains, as in reaction (2) (ref. 9), is considered to be a new kind of crosslink formation in the case of mechanical destruction.

We have observed changes in crosslink densities with time by chemorheological methods of forced extension of cured rubbers at low temperature. The samples used were natural rubber (NR) and nitryl rubber (NBR). The curing conditions are shown in Table 4.1. The cured samples were extracted with hot acetone for 48 h with a Soxhlet-type extractor, then maintained in a refrigerator after one day of drying under vacuum.

TABLE 4.1

Mixing detail and curing conditions

Sample	Rubber, parts	ZnO, phr	St-COOH,[a] phr	S, phr	M,[b] phr	DM,[c] phr	Curing temp., °C	Curing time, min
NR	100	5.0	1.0	2.0		1.0	145	10, 20, 30 40, 80
NBR	100	5.0	1.0	2.5	1.0		150	30, 40, 80

a Stearic acid.
b Mercaptobenzothiazole.
c Dibenzothiazyl disulfide.

Operation of an Autograph (IM-100) at a speed of 2 mm per min to extend dumbell-type specimens of 0.6 mm thickness, 5 mm width, and 20 mm length was carried out after preextension to 21 mm length in order to avoid errors from estimating the length of specimens after cracking. Forced strains were then applied on the materials after setting for 10 min at -78°C.

We used the normal intermittent stress-relaxation technique described above to measure the changes in crosslink density of sampl in a nitrogen atmosphere. To avoid shrinkage of samples under stress, the first recording of stress was made at each temperature, after extension for 45 min.

The stress-strain curve shown in Fig. 4.10 was the resultant of forces on extended rubber at low temperature. The maximum stress is found in a narrow range of α values (about 1.05). While the stress is decreasing, α is increasing; eventually the stress becomes constant and parallel to the α axis when α is more than 1.5. This corresponds to the sliding motions of rubber molecules which are strained beyond the yield point. This forced extensional ratio can be performed to about $\alpha=2$.

With change in extensional velocity, there was also a change in the elongation at break, and the lower the deformation, the larger the elongation at break.

Gaseous hydrogen molecules, which were freed from ruptured molecul have been observed on the surface of deformed specimens.

Fig. 4.11 was obtained at -78°C with an extensional velocity of

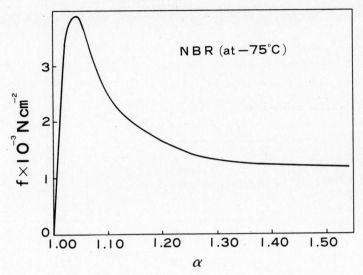

Fig. 4.10.Stress-strain curve obtained by forced extension of nitryl rubber (NBR) at low temperatures.

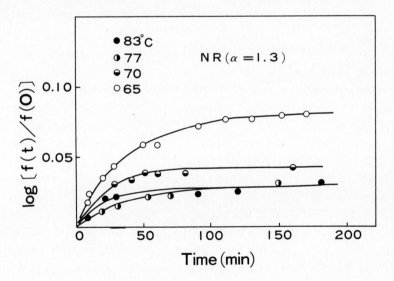

Fig. 4.11.Intermittent stress-relaxation curves for natural rubber (NR).

2 mm/min to extend the NR specimens to $\alpha=1.3$, then using intermittent stress-relaxation to measure changes after 45 min standing at each temperature (66°C, 70°C, and 83°C). It was shown that the lower the temperature, the larger the relative stress, and vice versa. The higher the temperature, the faster the recombination of cleft chains. In the preheated specimens, significant quantities of active radicals

130

capable of recombination are generated if the temperature is sufficiently low.

The same experiments were carried out for NBR and the results
showed the same tendency as those with NR. As can be seen from Fig.
4.12, the free radicals produced by mechanical destruction had com-
parable heat stability with NR and NBR.

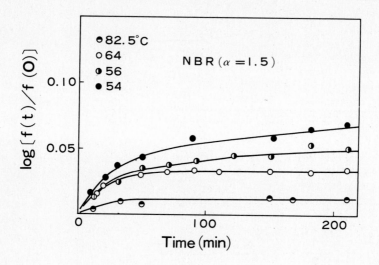

Fig. 4.12.Intermittent stress-relaxation curves for nitryl rubber (NBR).

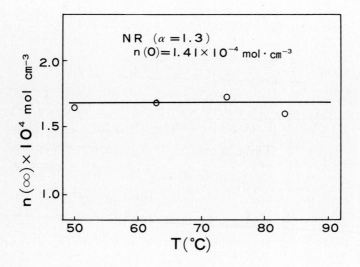

Fig. 4.13.Equilibrium crosslinking density n(∞) vs. temperature for natural
rubber.

From the plots of relative stress versus time in Figs. 4.11 and 4.12, it was found that the relative stress measured by the intermittent method increased rapidly at first and then approached a constant value gradually, a value corresponding to equilibrium. That is, after a certain time has elapsed, no more radicals appear to contribute to new crosslinking reactions.

In Fig. 4.13, the relation between equilibrium network $n(\infty)$ and temperature is plotted. $n(\infty)$ was obtained for samples left for 3 h at each temperature. The figure suggests that those radicals induced

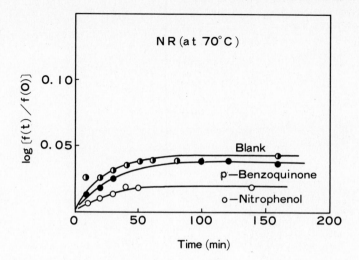

Fig. 4.14. Intermittent stress-relaxation curves for natural rubber which contains radical acceptors.

by chain-scission recombine at a constant ratio, regardless of the temperature. That is to say, the number of unstable radicals induced by constant extension is independent of temperature and is nearly constant. To further investigate these recombination reactions (refs. 6, 10), cured natural rubbers were immersed in benzene solutions containing the radical acceptors p-benzoquinone, 3.47 mg, and o-nitrophenol, 0.48 g per gram of NR.

Then, the specimens were dried in vacuo for one day after keeping them in the laboratory for 0.5 h. The same experiments as in Fig. 4.12 were carried out with these samples, and the results shown in Fig. 4.14 were obtained. In the presence of the radical acceptors, the results measured by intermittent stress relaxation showed a small increase in relative stress compared to the control specimens. This was due to the formation of reducing radicals which are the products

132

of reaction with radical acceptors; thus, the increasing networks
resulting from recombination of cleft chains are also considered to
be reduced subsequently. These experiments were carried out at constant
extension ratio.

Next, the equilibrium densities are discussed for n (∞), which we
measured at each temperature. With the same specimens mentioned above
the stress was measured by the same methods in a nitrogen stream at
70°C for 4 h after extension at low temperature.

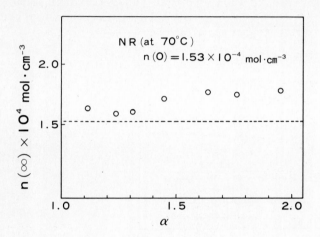

Fig. 4.15. Equilibrium crosslinking density n (∞) vs. extension ratio α for
natural rubber.

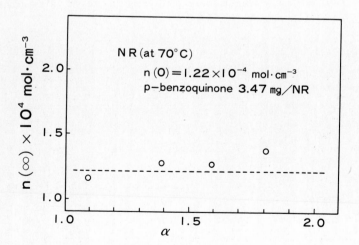

Fig. 4.16. Equilibrium crosslinking density n (∞) vs. extension ratio for natural
rubber which contains radical acceptors.

Assuming that the crosslinking density calculated from eqn
(2.1) is the equilibrium density, Fig. 4.15, in which radical acceptors were not included, and Fig. 4.16 in which radical acceptors were
included, were obtained. The broken lines shown in both figures indicate the initial crosslinking density of the specimens before molecular destruction. For the purpose of comparing these figures, Fig.
4.17 was obtained by plotting the ratio $n(\infty)/n(0)$ vs. α. It was shown
that, in the presence of radical acceptors, the ratio $n(\infty)/n(0)$ was
smaller than that without radical acceptors. Fig. 4.17 depicts the
relation between $n(1/2)/n(0)$ and shows that α approached the Maxwellian
curves, independent of $n(0)$, for three kinds of specimens having

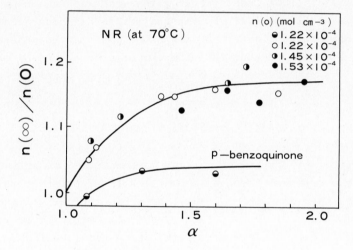

Fig. 4.17. Plot of $n(\infty)/n(0)$ vs. α for natural rubber.

different initial crosslinking densities. That is, molecular rupture,
which was induced by forced extension, depends on the extension ratio
α and is described by the appropriate Maxwellian curves. Thus the
change in properties induced by mechanical stimulus depends closely
on the strain ratio. These phenomena have been found not only in
molecular rupture at low temperature but also in the experiments at
large deformation involving stress relaxation of cured rubbers at
temperatures above room temperature (ref. 11).

The relation between $f(\infty)/f(0)$ and time can be described by the
Maxwellian relationship in the ordinary stress-relaxation curves; in
the case of molecular rupture induced by forced extension at low
temperature, as shown in Fig. 4.17, the same type of curves are
obtained by substituting time for the α axis.

The time measured for stress relaxation within the linear region

of strain behavior can be considered to be equivalent to the strain
ratio of large deformations. In Fig. 4.17, the increasing ratio of
crosslinking density in the wide range of α, as shown by Maxwellian
curves, becomes small. That is, as shown in Fig. 4.10, the mobility
of molecular chains has become large, depending on the magnitude of
extensional forces and its cleavage effect will then be minimal.

It is well known that the stability of free radicals induced by
mechanical shear differs from polymer to polymer. Poly (methyl
methacrylate) and polystyrene remain stable, even at comparatively hi
temperatures (refs. 12, 13) and the stability of these polymers is
closely related to their glass transition temperatures (ref. 14). Fro
the results measured by the intermittent stress-relaxation method
(as shown in Figs. 4.11 and 4.12) increasing stress has been maintained for a considerable time at the operating temperature. These
phenomena are believed to demonstrate the comparative stability of
radicals induced by the cleavage of chains from forced extension at
low temperatures.

III ABNORMAL BEHAVIOR IN CROSSLINKING REACTIONS

It is interesting and important to know whether a new crosslink
contributes to the observed force in a sample held at constant strain
A number of investigations (refs. 15-21), both theoretical and experimental, have been carried out on this subject. According to most
results, it appears that the new crosslinks which occur during
chemical relaxation do not appreciably contribute to the observed
relative stress in a sample held at constant elongation. However,
recent results obtained by us (refs. 22, 23) reveal a remarkable effec
of the additional crosslinkages on continuous chemical stress relaxation of crosslinked polyester polymer in air, and this is described
below.

The crosslinked polyester samples used here were prepared by
esterifying maleic anhydride with ethylene glycol and crosslinking
with divinylbenzene according to the usual method. The purpose of ou
experiment (refs. 22, 23) was to examine the effects of temperature,
elongation, and initial chain-density on a sample in which additional
crosslinkages occur during continuous chemical stress relaxation.

Fig. 4.18 shows the initial increased continuous chemical stress-
relaxation curve for samples having approximately the same value of
initial chain-density n(O) at constant strain ε. It was found from
Fig. 4.18 that the peak of the chemical relaxation curves becomes

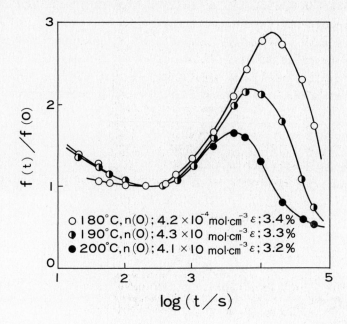

Fig. 4.18. Effect of temperature on the continuous chemical stress-relaxation curves in air.

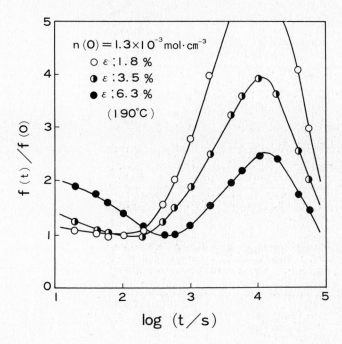

Fig. 4.19. Effect of strain ε on the continuous chemical-stress relaxation curves in air.

lower with increasing temperature. It is seen from Fig. 4.19 that the peak of the continuous stress-relaxation curve is reciprocally proportional to the strain ε for given n(O)=1.3 x 10^{-3} mol cm^{-3} and at a constant temperature of 190°C.

Similarly, it can be seen from Fig. 4.20 that the peak of the continuous chemical stress-relaxation curve is reciprocally propor-

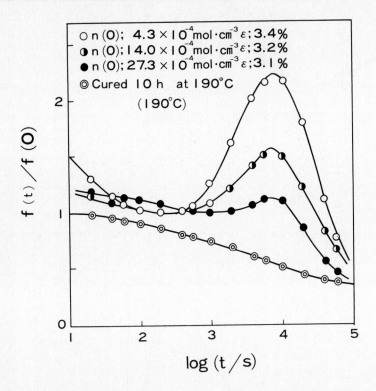

Fig. 4.20. Effect of initial density n(O) on the continuous chemical stress-relaxation curves in air.

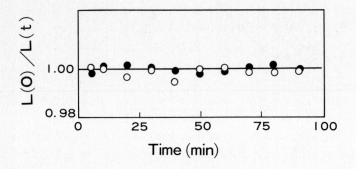

Fig. 4.21. The change of the original length with time at 190°C in air.

tional to the initial density n(O) at constant ε and temperature. The sample cured at 190°C for 10 h seems to be virtually completely cured and exhibits no upward curve in this figure. This shows that the unusual additional crosslinkages may contribute to the enhanced character of the chemical relaxation curve under continuous stress.

Although the change of the length L(t) of a sample with time was often measured under the various conditions mentioned above, no difference was observed, as shown in Fig. 4.21. This means that the stress-increase is not a consequence of sample shrinkage.

Tobolsky et al. recently also observed an obvious stress-increase in the absence of sample shrinkage in continuous stress-relaxation measurements on cis-polybutadiene under vacuum, but did not see the same effect in air (ref. 24).

The samples were cured cis-polybutadiene (Budene 501) supplied by the Goodyear Tire and Rubber Company; details are shown in Table 4.2.

TABLE 4.2

Composition and curing conditions

Component	Budene 501	Dicumyl peroxide, phr	Curing condition
Sample A	100	2.0	40 min at 135°C
Sample B	100	0.6	90 min at 135°C

All samples were subjected to thorough Soxhlet extraction in cyclohexane and acetone at their boiling temperatures for 48 h (ref. 25) after vulcanization, to remove chemical residues that might act as crosslinking agents. Ninety minutes was the minimum acceptable curing time for sample B; with shorter curing time the samples produced were appreciably soluble in cyclohexane benzene.

Stress-relaxation measurements were conducted both in air and in vacuo (0.1 Pa). The initial density of network chains n(O) for samples A and B at 230°C in vacuo were about 2.8-3.0 x 10^{-5} mol cm^{-3} and 2.4-2.6 x 10^{-4} mol cm^{-3}, respectively.

Fig. 4.22 shows continuous stress-relaxation curves for the relative stress $f_c(t)/f_c(O)$ vs. time for sample A in vacuo over the temperature range 200 to 260°C. The observed stress-decay is slower than expected from some previous papers. After the relative stress becomes about 0.85 to 0.90, the stress increases with temperature. After each run of measurements, the length of the sample was checked after returning to zero stress and no shrinkage could be observed within experimental

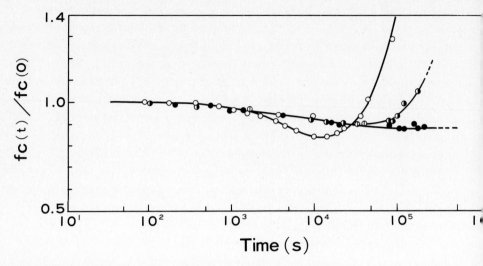

Fig. 4.22. Continuous stress-relaxation curves for sample A in vacuo at 200°C (●, α=1.05), 230°C (◐, α=1.05), and 260°C (○, α=1.05).

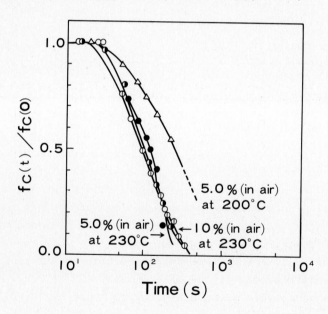

Fig. 4.23. Continuous stress-relaxation curves for sample A in air at 200°C and 230°C.

error (within 0.1%, about 0.1 mm). Fig. 4.23 shows results for sampl
A in air. No stress-increase was observed. Fig. 4.24 shows the resul
of continuous and intermittent stress-relaxation measurements.

The figure also includes results from the SMCIR method, described

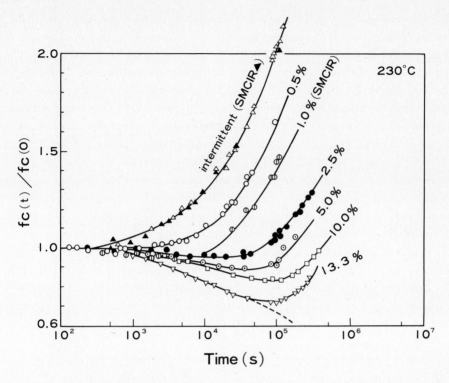

Fig. 4.24. Intermittent stress-relaxation curves and continuous stress-relaxation curves at various extension ratios for sample A in vacuo at 230°C.

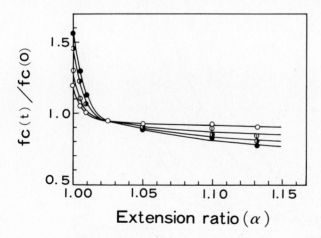

Fig. 4.25. The dependence of the relative stress on extension ratio for continuous stress-relaxation measurements in vacuum at 230°C as a function of time: ○, 500 s; ◐, 1000 s; ◑, 2000 s; ●, 3000 s.

in section 4.1.3. for sample A in vacuo at 230°C. In this case, the first extension ratio (α_1) was 1.01 and the second (α_2) was 1.05. The stress increases steadily in intermittent measurements. In Fig. 4.2

the stress curve for intermittent measurements is in good agreement

with that for intermittent SMCIR at α_1=1.01. It is known that the

stress-increase in intermittent measurements is caused by crosslinki

so the agreement between the results from two different intermittent

measurements means that the facility of the crosslinking reaction do

not depend on whether or not the sample is stretched. Fig. 4.24 also

shows that the stress in continuous measurements increases after a

certain period and that, with decreasing α, these curves approach th

in intermittent measurements. These facts show that the stress-incre

is caused by crosslinking even in the stretched sample and that the

contribution of crosslinking to the stress depends on α. Fig. 4.25

shows plots of relative stress at arbitrary times in Fig. 4.24 again

α. The relative stress seems to level off at α larger than 1.05. Th

smaller increase of the relative stress for larger α in Fig. 4.24 an

the levelling-off behavior for the relative stress in Fig. 4.25

suggest that the stress-increase in the continuous stress-relaxation

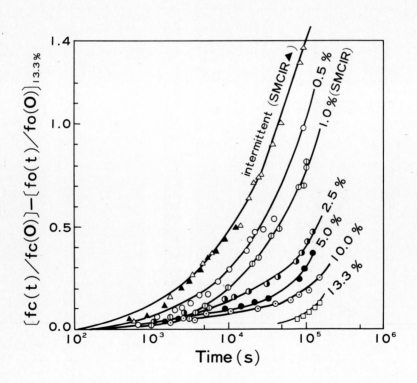

Fig. 4.26. $\{[F_C(t)/F_C(0)]_\alpha - [F_C(t)/F_C(0)]_\infty\}$ curves for sample A as a function of extension ratios obtained from Fig. 4.24. (Symbols as for Fig. 4.24.).

measurements disappears at large α. The dotted curve in Fig. 4.24 is drawn assuming that the stress-increase disappears. We denote this relative stress as $\{f_c(t)/f_c(0)\}_\infty$ based on the concept of "two-network theory" (ref. 27).

The magnitude of the stress-increase at an arbitrary extension ratio α can be given by the difference in the relative stress at α and ∞, $\{f_c(t)/f_c(0)\}_\alpha$, and $\{f_c(t)/f_c(0)\}_\infty$. The value of $\{f_c(t)/f_c(0)\}_\alpha - \{f_c(t)/f_c(0)\}_\infty$ for sample A in vacuo at 230°C is shown as a function of time for each extension ratio in Fig. 4.26. If we multiply by a suitable parameter $1/\beta$ adjusted for each curve in Fig. 4.26, we can superpose all of the curves on the intermittent curve, as shown in Fig. 4.27. The factor $1/\beta$ depends on α but not on time. In this procedure, β is expressed by the following equation.

$$\beta = \frac{\{f_c(t)/f_c(0)\}_\alpha - \{f_c(t)/f_c(0)\}}{\{f_i(t)/f_i(0)\} - \{f_c(t)/f_c(0)\}_\infty} \tag{4.19}$$

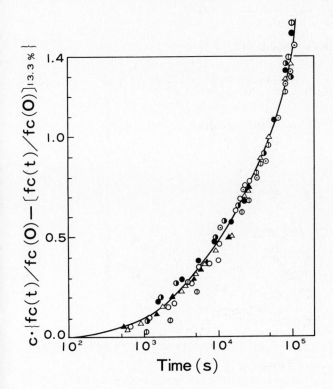

Fig. 4.27. Superposed curves of $1/\beta\{[F_c(t)/F_c(0)]_\alpha - [F_c(t)/F_c(0)]_\infty\}$ for sample A in vacuo at 230°C. (Symbols as for Fig. 4.24.).

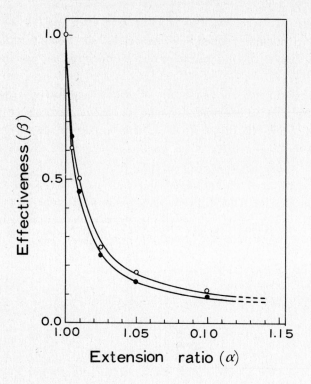

Fig. 4.28.Effectiveness β vs. extension ratio α for samples A and B in vacuo at 230°C: O, sample A; ●, sample B.

(a) Crosslinking of a free chain

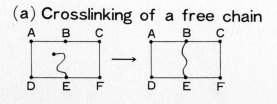

(b) Crosslinking of network chains

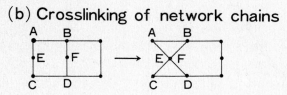

Fig. 4.29. Classification of crosslinking reactions in network structures.

where suffixes c and i correspond to continuous and intermittent stress-relaxation, respectively.

The denominator of eqn (4.19) shows the contribution of crosslinking to the stress-increase at zero extension, and the numerator is that at an extension of α. Therefore, β can be regarded as the fractional contribution of crosslinks to the stress-increase at α. We propose to call it the "effectiveness".

Plots of the effectiveness β vs. α are shown in Fig. 4.28. Here, we use the relative stress at $\alpha=1.133$ in place of $\{f_c(t)/f_c(0)\}_\infty$ in the light of the above discussions because $\alpha=1.133$ is the maximum extension ratio which could be achieved without breakage of samples. The effectiveness rapidly decreases in the region of small α and slowly approaches zero for large α. The dependence on α of the effectiveness β becomes smaller above $\alpha=1.10$. This fact also supports the use of the relative stress at $\alpha=1.133$ in place of $\{f_c(t)/f_c(0)\}_\infty$.

Crosslink formation in the stretched sample can be classified roughly into two groups, as shown in Fig. 4.29.

(a) Crosslinking of free chains in the network structure.

(b) Crosslinking of stretched network chains.

In the case of crosslinking of type (a) in Fig. 4.29, the chain BE in this figure is always in thermal equilibrium and not in extension. Therefore, crosslinking (a) contributes nothing to the stress in continuous stress-relaxation measurements, as the "two-network theory" indicates.

In the case of crosslinking of type (b), two network chains will be brought together. As a result of crosslinking reactions, two network chains are made to form four network chains. Each network chain will be extended to various extension ratios; some of them are larger and others smaller than α, but the average extension ratio for these network chains in the direction of the stress may be equal to α, according to the network theory of rubber; this is the reason that shrinkage of the sample could not be observed. The stress-increase is explained by the increase in the number of network chains formed by crosslinking of type (b).

We denote the number of increased network chains formed by crosslinking (a) as $2\varDelta C_a(t)$ and that by crosslinking (b) as $2\varDelta C_b(t)$. Then, the stress in continuous stress-relaxation measurements in which only crosslinking (b) contributes to the stress is written as follows,

$$f_c(t) = \{\, n(0) - q(t) + 2\,\varDelta C_b(t)\,\}\, RT(\alpha - \alpha^{-2}) \qquad (4.20)$$

On the other hand, the stress in intermittent stress-relaxation measurements in which crosslinking (a) and (b) both contribute to the stress increase is

$$f_i(t) = \{n(0) - q(t) + 2 \, \Delta C_a(t) + 2 \, \Delta C_b(t)\} \; RT(\alpha - \alpha^{-2}) \tag{4.21}$$

When the crosslinking does not contribute to the stress-increase, the stress is

$$f_c(t) = \{n(0) - q(t)\} \; RT(\alpha - \alpha^{-2}) \tag{4.22}$$

Substituting eqns (4.20), (4.21), and (4.22) into eqn (4.19), β is given by eqn (4.23):

$$\beta = \frac{\Delta C_b(t)}{\Delta C_a(t) + \Delta C_b(t)} \tag{4.23}$$

Equation (4.23) shows that the effectiveness β denotes the fraction of network chains formed by crosslinking (b). β does not show time-dependence, as seen in Fig. 4.28, and this behavior is explained easily using eqn (4.23) with the concepts of crosslinking reactions. Cross-linking reagents, such as dicumyl peroxide, react with double bonds to make crosslinks. There will exist no differences between two double bonds, one on a free chain and another on a network chain. This means that crosslinking reactions (a) and (b) have the same mechanism and that the time-dependence of both types of crosslinking reaction is the same.

In view of these considerations, it will be apparent that the time-dependence in eqn (4.23) can be cancelled out of the numerator and the denominator and for this reason β shows no time-dependence. These facts also mean that the fraction of network chains formed by crosslinking (b) is controlled only by the extension ratio α. Using the dependence of α on β, it will be possible to analyse the dependence on α of the reactivity of functional groups, and to control the number of network chains which have a certain direction. This is an interesting and important problem.

β is obtained from continuous stress-relaxation measurements, as seen previously. But, in air, it is impossible to make precise observation on the stress-increase, as in Fig. 4.23, even if cross-linking (b) occurs, because the stress decays at high velocity owing to oxidative scissions. Moreover, in air, the velocity of the stress-decay in continuous stress-relaxation measurements is restrained by the contribution to the stress of crosslinking (b).

If continuous stress-relaxation measurements are in high vacuum, repressing oxidative stress-decay, for the usual natural and synthetic rubbers, the stress-increase will be observed in the region of small α, as reported here for *cis*-polybutadiene.

Flory et al. (refs. 15, 18) reported that crosslinking reactions repress the stress-decay in continuous stress-relaxation measurements but, in practice, not merely repression of the stress-decay but actual stress-increase was seen. This means that their theory is applicable only to systems in which scission reactions occur very actively or at large extension ratios and that such reactions are not sufficient to explain the contribution made to the stress by crosslinking reactions in the stretched sample.

Usually, continuous stress-relaxation measurements are performed at large extension ratios, larger than 1.2. In such cases, the effectiveness β is very small, as presumed from Fig. 4.28, and this also means that the stress-increase becomes negligibly small. The "two-network theory" (ref. 27), therefore, assumes only an approximate validity.

At present, we are observing in our laboratories (ref. 26) interesting abnormal behavior similar to the stress-increase seen during continuous chemical stress relaxation caused by shrinkage of a sample on degradation; this differs essentially from the crosslinking reactions mentioned above.

Polyisobutylene oxide (PIBO), a typical photodegradable polymer,

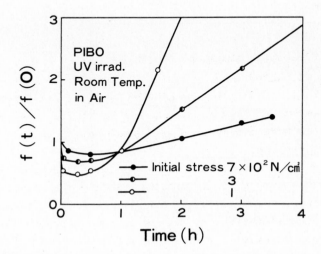

Fig. 4.30. Dependence of the strains on the relative stress for PIBO in air under irradiation.

was exposed to Ultra-Violet irradiation (250 watt Ushio Ultra-high pressure Hg lamp). It is clear from Figs 4.30 and 4.31 that the continuous chemical stress-relaxation curves of PIBO rise with photo-degradation while usual amorphous and crystalline polymers show the decay curves as degradation proceeds.

As shown in Fig. 4.32, the results of thermal oxidation without

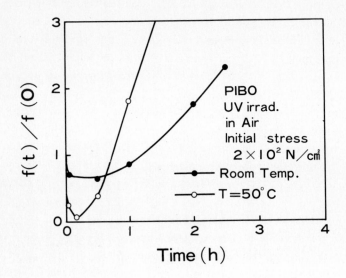

Fig. 4.31. Dependence on temperatures of the relative stress for PIBO in air under irradiation.

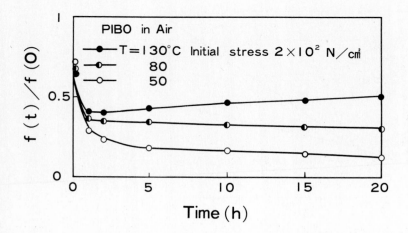

Fig. 4.32. Dependence on temperatures of the relative stress for PIBO in air without irradiation.

irradiation indicate that the relative stress increases gradually from 130°C and the degree of increase in air without U. V. irradiation is generally lower than that with U. V. irradiation. With the tendency shown in Fig. 4.33, that the relative stress increases very little in

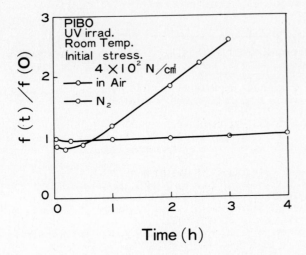

Fig. 4.33. Dependence on atmosphere of the relative stress for PIBO under irradiation in air and in N_2.

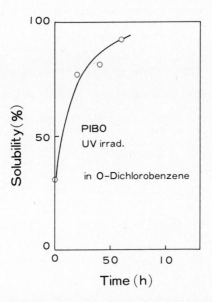

Fig. 4.34. The relation of solubility of PIBO in o-dichlorobenzene as function of irradiation time.

148

N_2 even if exposed to irradiation, oxidative degradation is seen to be responsible for the stress-increase in the case of PIBO.

Since actual shrinkage of the sample was observed, and a rapid increase of solubility of PIBO in o-dichlorobenzene with irradiation (as shown in Fig. 4.34) was observed, the abnormal behavior for PIBO is attributed to shrinkage resulting from oxidative degradation.

REFERENCES

1 A.V. Tobolsky, I.B. Prettyman, and J.H. Dillon, J. Appl. Phys., 15(1944)380.
2 R.D. Andrews, A.V. Tobolsky, and E.E. Hanson, J. Appl. Phys., 17(1946)352.
3 S. Ore, J. Appl. Polym. Sci., 2(1959)318.
4 H. Sobue, K. Matsuzaki, T. Migita, Y. Mitsuda, and K. Murakami, Kobunshi Kagaku (Polymer Chemistry), Japanese Ed., 21(1964)606.
5 K. Goto, Kogyokagaku Zasshi, 71(1968)1319.
6 S. Yamashita, Kogyokagaku Zasshi, 71(1968)1331.
7 A.A. Berlin, G.S. Petrov, and V.F. Prosvirkina, Zhr. Fiz. Khim., 32(1958)2565.
8 K. Shinohara and K. Kashiwabara, Hoshasen to Kobunshi, Japanese Ed., Makishoten 1968.
9 E.H. Andrews and P.E. Reed, Poly. Lett., 5(1967)317.
10 G. Ayrey, C.G. Moor, and W.F. Watson, J. Polym. Sci., 14(1956)1.
11 T. Kusano and K. Murakami, paper presented at 18th Annual Meeting of the Society of Polymer Science of Japan, Kyoto, 1969; T. Kusano, Y. Suzuki, and K. Murakami, J. Appl. Polym. Sci., 15(1971)2453.
12 P.Y. Butyagin, I.V. Kolbarev, A.M. Dubinskaya, and M.V. Kisluk, Vysokomol. Soedin., 10(1968)2265.
13 S.E. Bresler, S.N. Zhurkov, S.N. Kazbekov, E.M. Saminski, and E.E. Tomashevski, zhur, Tekh. Fiz., 29(1959)358.
14 D.K. Backman and K.L. Devries, J. Polym. Sci., A-1, 7(1969)2125.
15 P.J. Flory, Trans. Faraday Soc., 56(1960)722.
16 J. Scanlan, Trans. Faraday Soc., 57(1961)839.
17 D.K. Thomas, Polym. 7(1966)125.
18 J. Berry, W.F. Watson, and J. Scanlan, Trans. Faraday Soc., 52(1956)1137.
19 K. Murakami et al., J. Soc. Materials Sci., Japan, 15(1966)312.
20 K. Murakami, T. Kusano, S. Naganuma, and Y. Takahashi, Bull. Chem. Research Institute of Non-aqueous Solutions, Tohoku Univ., Japan, 19(1969)243.
21 S. Naganuma and Y. Takahashi, Kobunshi Kagaku (Chem. High Polymers, Japan), 27(1970)705.
22 K. Murakami, T. Kusano, S. Naganuma, and Y. Takahashi, Kogyo Kagaku Zasshi, 74(1971)1439.
23 K. Murakami and S. Tamura, J. Polym. Sci., Polym. Lett. Ed., 10(1972)941.
24 A.V. Tobolsky, Y. Takahashi, and S. Naganuma, Polym. J., 3(1972)60.
25 J.R. Dunn, J. Appl. Polym. Sci., 7(1963)1543.
26 K. Murakami and M. Yoshinari, J. Polym. Sci., in press.
27 A. Tobolsky, and H. Eyring, J. Chem. Phys., 11(1943)125.

CHAPTER 5

APPLICATION OF DYNAMIC MECHANICAL METHODS TO CHEMORHEOLOGY

In the previous chapters, the stress-relaxation caused by chemical
reaction of crosslinked polymers has been the main subject of discussion. The stress measurements were carried out under constant
strain in these studies, and the technique may be regarded as a
"method of transients". Dynamic mechanical methods may equally be
applicable to the estimation of stress change and some aspects of
the application of dynamic methods to chemorheology will be surveyed
briefly in this chapter.

The study of the dynamic mechanical responses of polymers undergoing simultaneous reaction of network chains comprises two classes
of investigation. The first is concerned with the measurement of
dynamic modulus under small strains and the second with cyclic stress-
relaxation in large deformations. Since the mechanisms of stress-decay
in these two classes of investigation differ considerably, they are
described separately in sections 5.1 and 5.2.

I. DYNAMIC MECHANICAL PROPERTIES OF CROSSLINKED POLYMERS UNDERGOING
OXIDATION

Either physical relaxation or chemical reaction of network chains
is responsible for the dynamic mechanical response of crosslinked
polymers undergoing oxidation. First, we focus our attention on the
physical relaxation of non-reacting networks where oxidation is absent.

The dynamic mechanical responses for lightly crosslinked polymers,
such as rubber vulcanizates, has been investigated extensively by
Ferry et al. (ref. 1).

The complex shear-compliances have been measured over a wide frequency range. Using the method of reduced variables, all the data
were reduced to a uniform reference temperature. Anomalous low frequency losses were observed, especially for non-sulfur vulcanizates.
retardation spectra L, calculated from the reduced storage and loss
compliance for various vulcanizates of natural rubber, are shown in
Fig. 5.1 (ref. 2). According to Rouse theory, modified for networks
(ref. 3), L should vanish at long times. As shown in Fig. 5.1, most
of the non-sulfur vulcanizates deviate from this prediction. It was

150

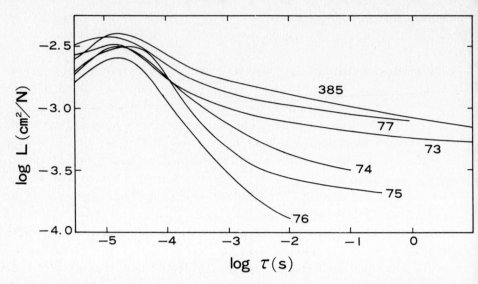

Fig. 5.1. Retardation spectra of natural rubber vulcanizates reduced to 25°C. Average molecular weight between crosslinks for 74, 75, 76 (sulfur cure), 73 (TMTD cure), 77 (DCP cure), and 385 (radiation cure) samples are 4,700, 4,300, 3,200, 6,400, 8,900, and 7,400, respectively.

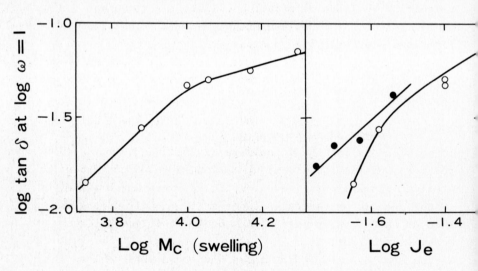

Fig. 5.2. Loss tangent at $\log \omega a_T = 1$, plotted logarithmically against molecular weight between crosslinks and equilibrium compliance for natural rubber vulcanizates. Open circles, DCP vulcanizates; black circles, sulfur vulcanizates.

also found that the magnitude of the loss tangents in this range increases with increasing spacing between crosslinks (Fig. 5.2, ref. 4). Evidently, the anomaly can be ascribed to some type of network defect of which three important types should be taken into account, i.e., the presence of loose ends due to the finite molecular weight of the primary polymer chains, the presence of coupling entanglements, permanently trapped in the course of crosslinking, and the presence of the sol fraction (ref. 1).

The relaxation times for loose ends would be too short to contribute to the low frequency losses shown in Fig. 5.1. It is also believed that permanently trapped entanglements do not have sufficiently long relaxation times to account for the low frequency dispersion. The only important mechanism for the low frequency losses is attributed to the relaxation due to entanglements which are involved in branched floating structures, partially attached to the gel network. Such ent- anglements may be referred to as "untrapped entanglements", and can have very long relaxation times.

The quantitative estimation of the numbers of trapped and untrapped entanglements is a difficult problem, since there appears to be no really reliable means to determine the number of elastically effective network chains. However, a simple criterion for entanglement entrapment (ref. 6) is successfully introduced in relating the equilibrium modulus to the pseudo-equilibrium one in the low frequency region. In the treatment of Kraus (ref. 7), any entanglement of a molecular chain running between two chemical crosslinks is trapped, as shown in Fig. 5.3a. In this case, the number of moles of elastically effective

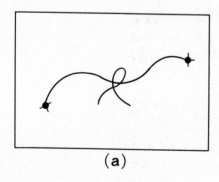 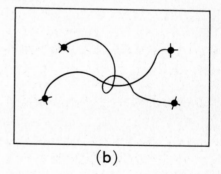

(a)   (b)

Fig. 5.3. Models for entanglement trapping. (a) Kraus (ref. 7), (b) Mancke, Dickie, and Ferry (ref. 6). Black circles are chemical crosslinks.

network chains ν is represented by,

$$\nu = 2(\nu_c + \varepsilon)(1 - 1/2\nu_c \bar{M}n) \tag{5.1}$$

where, ν_c is the moles of chemical crosslinks, b is a constant appro- ximately equal to twice the polymer density, ε is the number of moles of entanglements, and $\bar{M}n$ is the number average molecular weight for the primary polymer chains. In this model, the proportion of trapped entanglements T_e is $1 - b/2\,\nu_c\bar{M}n$, but this counts, too many trapped entanglements. A more realistic model was proposed (refs. 1, 6) in which all four network chains radiating from an entanglement must be terminated by chemical crosslinks, as shown in Fig. 5.3b. In this case, eqn (5.1) is rewritten as,

$$\nu = 2\nu_c(1 - b/2\,\nu_c\bar{M}n) + 2\,\varepsilon(1 - b/2\,\nu_c\bar{M}n)^2 \tag{5.2}$$

Here, the probability of entanglement entrapment is

$$T_e = (1 - b/2\nu_c\bar{M}n)^2 \tag{5.3}$$

In lightly crosslinked networks, the storage moduli and compliance show a plateau region, which may be attributed to the contributions from all the entanglement points plus the chemical crosslinks. An example is shown in Fig. 5.4 (ref. 8). According to the statistical theory of rubberlike elasticity, the plateau molulus G_{en} and plateau

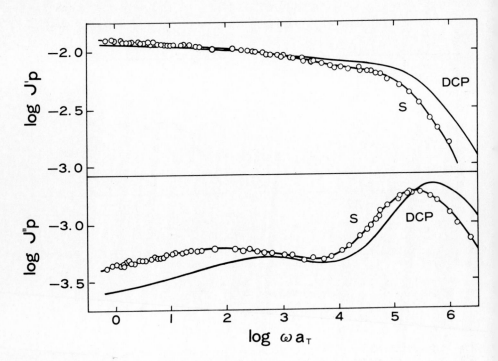

Fig. 5.4. Storage and loss compliances of natural rubber reduced to 25°C. Numbers denote $\nu \times 10^4$ mol cm^{-3}, where ν is the crosslink density.

compiiance J_{en} are represented by,

$$G_{en}/G_e = J_e/J_{en} = \nu_{en}/\nu \tag{5.4}$$

where G_e and J_e are the equilibrium modulus and compliance, and ν_{en} is the number of moles of network chains terminated by chemical crosslinks or by entanglements, whether trapped or not. ν_{en} is represented by,

$$\nu_{en} = 2\nu_c(1 - b/2\,\nu_c\bar{M}n) + 2\varepsilon \tag{5.5}$$

For the corresponding uncrosslinked polymer (ref. l),

$$J_{en}{}^o = 1/2g\ \varepsilon RT \tag{5.6}$$

Where g is the front factor, R is the gas constant and T is the

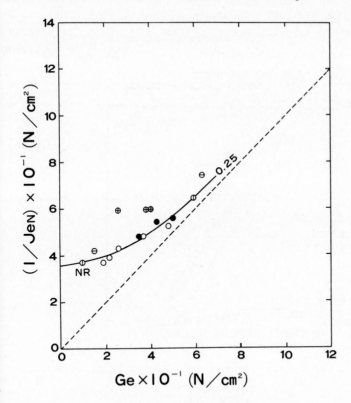

Fig. 5.5. Reciprocal of plateau compliance plotted against equilibrium modulus. Solid curve calculated with indicated value of $(bgRT/\bar{M}n) \times 10^{-6}$. Open circles, including those with horizontal and vertical plots, are DCP vulcanizates. Black and crossed circles denote sulfur and TMTD vulcanizates, respectively.

absolute temperature. The elimination of ν_c, ν_{en} and ε from eqns (5. (5.4), (5.5), and (5.6) gives a relation between G_e and G_{en}, if the quantity $bgRT/\bar{M}n$ is shown. The relation is tested for natural rubber vulcanizates by the experimental data in Fig. 5.5 (ref. 6).

The data are fitted well with an appropriate $bgRT/\bar{M}n$ value, irrespective of crosslink structure. The result seems to support the hypothesis that the low frequency loss mechanism is due primarily to the relaxation of untrapped entanglements.

Let us now consider a network where oxidative chain-scission occurs. As revealed in the foregoing discussions, the storage modulu at a given frequency involves contributions from the chemical cross-links and the untrapped entanglements which cannot rearrange within a period of oscillatory deformation. In the reacting system, either the number of chemical crosslinks or the number of untrapped entangl ments becomes a function of time. The effect of the untrapped entang ments might be minimized by extracting the sol fraction before meas-urement, however, the crosslinks or the entanglements which are elas cally active in the initial stage become elastically ineffective as the reaction proceeds and the presence of such chains would contribu to low frequency losses. Hence, physical relaxation cannot be

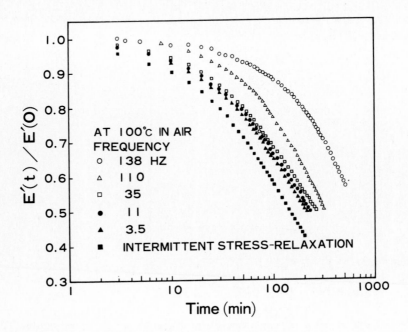

Fig. 5.6. Frequency dependence of the decay of relative relaxation modulus. The intermittent stress-relaxation curve is also plotted for comparison.

neglected, especially in the later stages, unless the relaxation times of the network chain fragments are short enough. Recent results of Yano and co-workers (ref. 9) clearly confirm the above prediction (Fig. 5.6).

The storage modulus at a given oxidation state increases with increasing frequency, indicating that contributions from network chain fragments produced in the oxidation are important. It is also found that the difference between the storage modulus at a given frequency and the relaxation modulus obtained by intermittent stress measurement increases as the oxidation proceeds.

It follows from the above discussion that simple additivity between the contributions of physical relaxation and network chain-scission cannot hold unless the relaxation times for both processes are far apart on the time scale. In order to separate the contributions of physical relaxation and network chain-scission, we must know the viscoelastic response of the polymer network at various oxidation states and the kinetics of oxidation (ref. 10). The former can only be estimated by dynamic mechanical method over a wide frequency range. For the investigation of kinetics of oxidation, the dynamic mechanical method seems to have no advantages over the conventional stress-relaxation experiment.

Another application of the dynamic mechanical method to the reacting network system is the investigation of the curing process of rubbers.

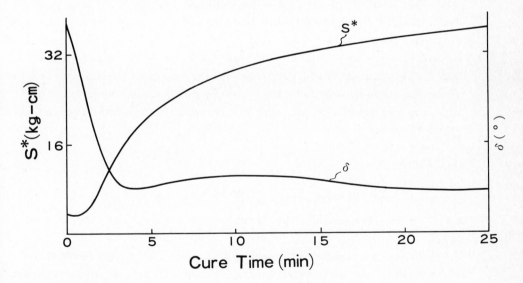

Fig. 5.7. Curing curve for DCP cured cis-polybutadiene S* denotes apparent torque value obtained by Curelastometer (ref. 11).

Many types of oscillating disc rheometer have been developed to obta
the optimum conditions for vulcanization. An interesting result of
Umeno and others (ref. 11) is shown in Fig. 5.7.

The loss angle δ attains a maximum value, as the vulcanization
proceeds, which approximately corresponds to the gel point, where th
high molecular chain fragments become attached to the network, and
then contribute to the low frequency loss mechanism.

The dynamic mechanical method appears to be of importance in maki
clearer the viscoelastic responses of crosslinked polymers undergoin
chemical reaction.

II DYNAMIC CHEMORHEOLOGY IN LARGE DEFORMATIONS

In certain practical conditions, network polymers are used under
oscillating deformation with large amplitudes. To obtain the necessa
technical information, various dynamic mechanical procedures have
been employed. In general, the analysis of the viscoelastic response
of such complicated systems is difficult for the following two reaso
First, the viscoelastic behavior is non-linear, when a large deforma
tion is imposed. Second, the primary chemical bonds are cut without
macroscopic fracture. Detailed discussion of such behavior is beyond
the scope of this book.

At high temperatures, the oxidative degradation of the network
chains of rubber vulcanizates is accelerated by the rupture of molecular chains due to mechanical force and the chemorheological studie
of such systems were undertaken by Kusano and Murakami (refs. 12, 13

Sulfur-cured natural rubber was used for these experiments. A
special apparatus (ref. 12) was designed to perform the stress-relax
ation measurements under large cyclic deformation by changing a
rotating motion to an oscillating one (Fig. 5.8). The applied strain
is of the form,

$$\gamma(t) = \frac{\gamma_{max}}{2}(1 + \sin 2\pi ft) \qquad (5.7)$$

where $\gamma(t)$ and γ_{max} are the strain at time t and the maximum strain,
and f is the frequency; the frequency can be varied from 0.7 to 225
cpm.

In order to determine the critical temperature for oxidative
degradation, the usual continuous-stress relaxation was carried out
for small strains and the slope k of the Maxwellian decay curves is
plotted against temperature (Fig. 5.9). Oxidative degradation become

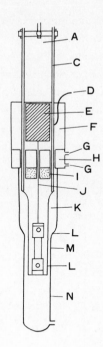

Fig. 5.8. Apparatus for stress-relaxation measurements under cyclic strain. A: For oscillatory motion; B: piston and guide; C, K: sliding bars; D, E, J: strain gauges; F: setting part of frequency and elongation; G: gear motor; H: cooler; I: cork; L: clamps; M: sample; N: glass tube for controlling atmosphere.

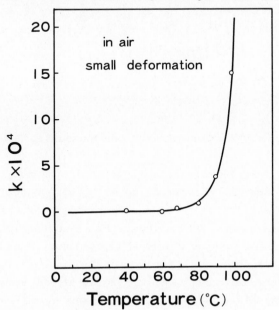

Fig. 5.9. Relation between the rate constant for stress-decay in air and temperature for small deformations.

158

significant above approximately 70°C in air.

On applying the strain, represented by eqn (5.7), the response of
stress is no longer represented by the trigonometric function, owing
to the non-linear viscoelastic behavior of the rubber vulcanizates.
As an approximation, the maximum stress at each cycle was taken as a
representative stress value specifying the relaxation of the stress
with time. A typical example at 88°C is shown in Fig. 5.10. Under
constant elongation, the stress-relaxation curves under cyclic strain

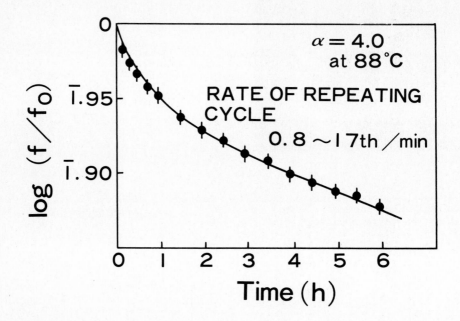

Fig. 5.10. Stress-relaxation under cyclic strain. Relaxation curve is independent
of frequency.

obtained at various frequencies, coincide with one another within
experimental error.

In Figs. 5.11 and 5.12, similar stress-relaxation curves for
various elongations are compared. The curves coincide with one
another for small strains, while the rate of stress-decay increases
with increasing strain for large strains. These results are quite
analogous to the behavior observed in the usual continuous stress-
relaxation experiments, that is to say, regions of linear behavior
would also exist in stress-relaxation under cyclic strain at a given
temperature. The energy of activation E_a of the stress-relaxation is
shown in Fig. 5.13 as a function of extension ratio α. In the region
$\alpha < 1.5$ and $\alpha > 3$, the energy of activation has a nearly constant

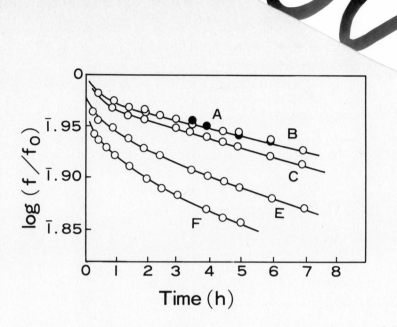

Fig. 5.11. The effect of extension ratio on the rate of stress–decay under cyclic strain, measured at 88°C in air. Frequency: 14 cpm.
A: $\alpha=1.25$, B: $\alpha=1.5$, C: $\alpha=2.5$, D: $\alpha=3.25$, E: $\alpha=4.0$, F: $\alpha=5.0$.

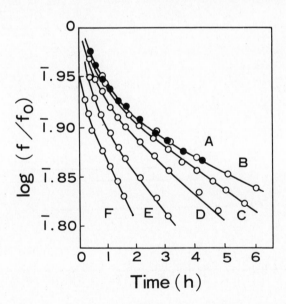

Fig. 5.12. The effect of extension ratio on the rate of stress–decay under cyclic strain, measured at 100°C in air. Frequency: 14 cpm.
Symbols are the same as in Fig. 5.11.

value, while a steep increase with increasing α is observed in the intermediate region. The constant value of E_a at higher elongations is probably ascribed to an orientational effect of the network chains.

160

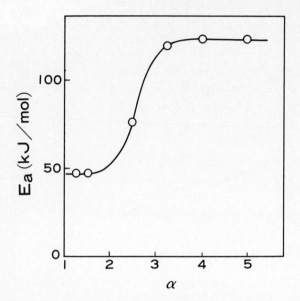

Fig. 5.13. Plot of activation energy against extension ratio.

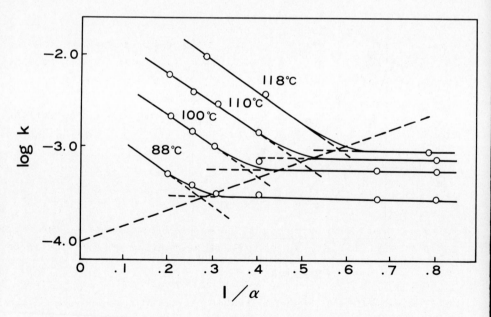

Fig. 5.14. Relation between log k and 1/α at 14 cpm.

As can be seen in Figs. 5.11 and 5.12, at a constant frequency, the rate of stress-decay k increases with increasing elongation at high elongations. Thus, the increase of α has similar influence upon the rate of stress-decay as does increase of temperature. In Fig. 5.14, the rate of stress-decay is plotted as a function of reciprocal exten- sion ratic 1/α at various temperatures. For small deformations, straight lines parallel to abscissa are obtained, indicating linear relaxation behavior without any mechanical effect. On the other hand, the logarithmic rates of stress-decay for large deformations are represented by straight lines with a slope -3.0, independent of temperature.

These results suggest that the mixed effects of chemical and mechanical degradation arise above a certain extension ratio or above a critical temperature, and that these effects are independent of each other. This makes it possible to estimate quantitatively the mechano-chemical effects on rubber networks.

When the oxidative scission of network chains proceeds according to first order kinetics, the chemical stress-decay is represented by (ref. 14),

$$f(t)/f(0) = e^{-kt} \tag{5.8}$$

where $f(t)/f(0)$ is the relative stress. As shown in Fig. 5.14, the rate constant k is independent of α for small deformations, while it is a function of α for large deformations. It was found that the stress-relaxation curves under cyclic strain are represented by eqn (5.8), for both small and large deformations.

Let us limit our attention to the region of large deformations. From Fig. 5.14, the rate of stress-decay $k(\alpha,T)$ in this region is given by,

$$\ln k(\alpha,T) = -\frac{E_f}{\alpha} + \ln A_f(T) \tag{5.9}$$

where, E_f and $A_f(T)$ are the slopes and the intercepts of the straight lines in Fig. 5.14. E_f specifies the mechanical effects for large deformations.

Now, the temperature-dependence of stress-decay k is represented by the following Arrhenius-type eqn:

$$k(\alpha,T) = A(\alpha) \cdot e^{-(E/RT)} \tag{5.10}$$

where, E is the energy of activation and $A(\alpha)$ is the pre-exponential factor, which is a function of the extension ratio α. From eqns

162

(5.9) and (5.10) we obtain,

$$\ln\frac{k(\alpha,T)}{k(\alpha_0,T_0)} = E_f(\frac{1}{\alpha_0} - \frac{1}{\alpha}) + \frac{E}{R}(\frac{1}{T_0} - \frac{1}{T}) \qquad (5.11)$$

where α_0 and T_0 are a reference extension ratio and a reference temperature, respectively. Eqn (5.11) indicates that the complicated dependence of the stress-decay curves on both mechanical and chemical effects may be separated as a function of extension ratio alone and a function of temperature alone. Any desired rate of stress-decay for large deformations can be calculated from eqn (5.11), provided that the constants E_f and E are known.

As shown in Fig. 5.14, the critical extension ratio α_c for the transition between linear and non-linear mechano-chemical behavior increases with decreasing temperature. This is shown by a dashed line. The intercept of this line on the vertical axis represents a critical rate of stress-decay, below which the non-linear behavior vanishes, that is to say, the mechanical effect is absent of all extension ratios.

The critical extension ratio α_c is plotted against reciprocal absolute temperature in Fig. 5.15. The experimental points are

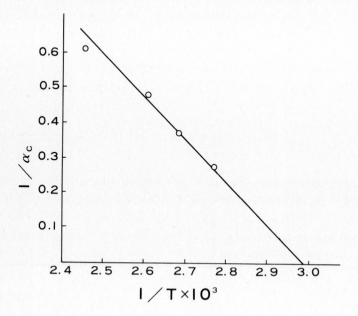

Fig. 5.15. Relation between $1/\alpha_c$ and $1/T$.

well fitted by a straight line. The intercept on the abscissa indicates the critical temperature where the mechanical effect vanishes. This is calculated to be 63°C. Therefore, degradation due to mechanical rupture of network chains should not occur below 63°C.

REFERENCES

1 J.D. Ferry, Viscoelastic Properties of Polymers, Wiley, New York, 2nd ed., 1970, Ch. 14, p.431.
2 R.A. Stratton and J.D. Ferry, J. Phys. Chem., 67(1963)2781.
3 M. Mooney, J. Polym. Sci., 34(1959)599.
4 J.D. Ferry, R.G. Mancke, E. Maekawa, Y. Oyanagi, and R.A. Dickie, J. Phys. Chem., 68(1964)3414.
5 N.R. Langley and J.D. Ferry, Macromolecules, 1(1968)353.
6 R.G. Mancke, R.A. Dickie, and J.D. Ferry, J. Polym. Sci., A2, 6(1968)1783.
7 G. Kraus, J. Appl. Polym. Sci., 7(1963) 1257.
8 E. Maekawa, R.G. Mancke, and J.D. Ferry, J. Phys. Chem., 69(1965)2811.
9 S. Yano, M. Murayama, S. Nakada, and M. Ono, Abstracts of 23rd Discussion Meeting of Rheology, Japan, 1975, p.93.
10 J. Moacanin, J.J. Aklonis, and R.F. Landel, J. Macromol. Sci. -Phys., B11(1974) 41.
11 M. Umeno, D. Thesis, Kyoto University, 1975.
12 T. Kusano and K. Murakami, J. Polym. Sci., A1, 10(1972)2823.
13 K. Murakami and T. Kusano, Rheol. Acta, 13(1974)127.
14 A.V. Tobolsky, Properties and Structure of Polymers, Wiley, New York, 1960, Ch. 5, p.225.

CHAPTER 6

APPLICATIONS OF CHEMORHEOLOGY

As described in the previous chapters, chemorheology affords a
simple and sensitive method for investigating the chemical changes
occurring in rubber networks. Although it does not give any direct
information about the chemical reaction itself, quantitative analysis
of chemical reactions becomes possible when a simple relation holds
between the amount of a chemical species present and a bulk mechanica
property. Therefore, it is especially of use for kinetic studies of
systems where the mechanical properties are profoundly influenced by
a chemical reaction. In this respect, vulcanized rubbers seem to be
most suitable for the application of chemorheology.

The chemorheological studies of rubber networks are based on the
theory of rubber elasticity. According to this theory, the modulus
of rubber is simply determined by the density of the network chains,
not by the density of the repeating units. Generally, network chains
of vulcanized rubbers consist of 50 or more repeating units, hence
the measurement of the modulus is expected to be at least 50 times as
sensitive as the measurement of other properties.

A similar situation, however, cannot be expected to hold for solid
polymers without chemical crosslinks. Furthermore, the physical
relaxation of polymer chains is not negligible in these cases. For
uncrosslinked polymers, a general treatment of relaxation accompanied
by chemical reaction seems extremely difficult, since the viscoelasti
response at various stages of reaction must be known. An example
using simplifying assumptions, will be discussed in section 1 of this
chapter.

Another limitation for chemorheology is encountered in the case of
the heterogeneous polymers. The rate of a chemical reaction is genera
different in various phases. This makes it difficult to correlate the
changes in mechanical properties with the chemical reaction. Further
more, the mechanism of stress-transfer is much more complicated in
these systems. In any case, detailed morphological investigations sho
be carried out before chemorheological methods are applied to heterogeneous systems, such as crystalline polymers or filled polymers.

In spite of the above limitations, chemorheology has several advantages in monitoring chemical reaction, especially in practical situations. First, the stress patterns employed in chemorheology are often close to those used in practical conditions. In some cases, the experimental data obtained in the laboratory serve directly to provide the technological information necessary for the materials. Second, the method is non-destructive.

This chapter is concerned with the applications of chemorheology to polymer technology. The applications to crosslinked rubbers are omitted because these are extensively discussed in the previous chapters.

I PLASTICS

1 Chemorheology of Linear Polymers

The chemorheology of linear polymers (uncrosslinked polymers) was difficult to treat theoretically because molecular flow by diffusion is generally much more rapid than relaxation or flow caused by chemical reaction. This section deals with chemorheology of linear polymers now that it has become possible to overcome the difficulties in some cases. While in crosslinked polymers there is some physical relaxation which can be neglected in a general discussion, both chemical and physical relaxation must be considered in the rubbery-flow region of uncrosslinked polymers.

Tobolsky et al. (ref. 1) derived eqn (6.1) from the box-shaped distribution of relaxation times with respect to the relaxation modulus $E_r(t)$ in the rubbery-flow region of amorphous polymers.

$$E_r(t) = E_0 \{E_i(- t/\tau_e) - E_i(- t/\tau_m)\} \tag{6.1}$$

Here, E_0 is the height of the box-shaped relaxation spectrum, τ_e and τ_m are the minimum and the maximum relaxation times, respectively, in the rubbery-flow range. $E_i(-x)$, which is the integral exponent, can be expanded by eqn (6.2),

$$E_i(- x) = - \int_x^{\infty} (e^{-u}/u)du = \ln x + \gamma - x +(- x)^r/r! + \cdots \tag{6.2}$$

where γ is the Euler constant, equal to 0.5772.

Assuming a first-order reaction with regard to the decomposition of the polymer, where k_1 is the rate constant and $M(O)$ and $M(t)$ are the molecular weights initially and at time t, respectively, eqns

(6.3) and (6.4) will be valid:

$$-dM(t)/dt = k_1 M(t) \qquad (6.3)$$

$$M(t) = M(0) \cdot e^{-k_1 t} \qquad (6.4)$$

It was confirmed that the rate of peroxide-accelerated degradation of polystyrene is well represented by these equations.

Now, it is known that τ_m is proportional to the 3.4th power of the molecular weights, hence initially and at time t, respectively, eqns (6.5) and (6.6) will be valid:

$$\tau_m(0) = k_2 \{M(0)\}^{3.4} \qquad (6.5)$$

$$\tau_m(t) = k_2 \{M(t)\}^{3.4} \qquad (6.6)$$

On substituting the ratio of $\tau_m(t)$ to $\tau_m(0)$ into eqn (6.4), eqn (6.7) is obtained:

$$\tau_m(t)/\tau_m(0) = e^{-3.4k_1 t} \qquad (6.7)$$

Again, on substituting the time-dependence of $\tau_m(t)$ for τ_m in eqn (6.1), the theoretical eqn (6.8) for the chemical degradation and physical relaxation for a linear amorphous polymer can be derived (ref. 2).

$$E_r(t) = E_o \{E_i(- t/\tau_e) - E_i(- t/\tau_m(0) \cdot e^{-3.4k_1 t})\} \qquad (6.8)$$

A monodisperse sample, an anionic-polymerized styrene, and two polydisperse samples were used in accelerated degradation experiments; the molecular weights of the samples are listed in Table 6.1.

To measure the chemical relaxation at 120°C, samples containing 2, 3, or 5 parts of dicumyl peroxide (DCP) were cast from dichloromethane solutions and dried in vacuo for 2 weeks. The relaxation modulus was measured by using the apparatus in our laboratories (ref. 3).

At various states of degradation, samples containing dicumyl peroxide were reprecipitated with benzene-methanol at 0°C, and their number-average and weight-average molecular weights were measured by gel-permeation chromatography.

TABLE 6.1

Molecular weight and heterogeneity index of molecular weight distribution for polystyrene

Polystyrene	$\bar{M}_n \times 10^{-6}$	$\bar{M}_w \times 10^{-6}$	$\bar{M}_w/\bar{M}_n$
Monodisperse	3.61	4.15	1.15
Polydisperse	3.24	6.31	1.95
Polystyrene PS-2[a]	0.08	0.28	3.26

a   Commercial samples from RAPRA, England.

The stress-relaxation curves for monodisperse and polydisperse polystyrenes are given in Figs. 6.1 and 6.2.

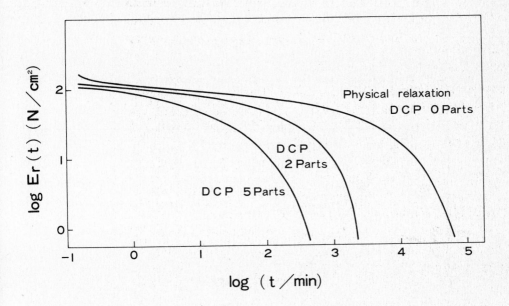

Fig. 6.1. Physical and chemical relaxation of monodisperse polystyrene at 120°C in N_2.

Master curves of physical relaxation are shown at the reference temperature 120°C. In samples containing dicumyl peroxide at 120°C in N_2, the stress decays faster with increasing dicumyl peroxide.

The experimental data for $-\ln\{M(t)/M(O)\}$ versus time are plotted in Fig. 6.3 and 6.4. As straight lines are obtained for both polystyrenes containing 5 parts dicumyl peroxide, the rate constants of decomposition are calculated as $K_1 = 2.10 \times 10^{-3}$ min^{-1} for monodisperse polystrene and as $K_1 = 2.04 \times 10^{-3}$ min^{-1} for polydisperse polystyrene.

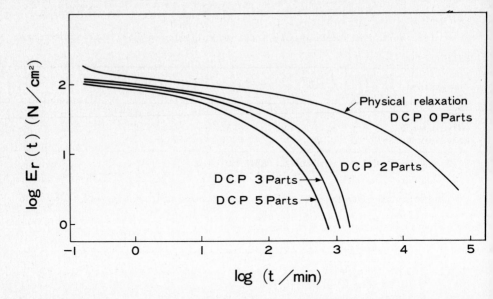

Fig. 6.2. Physical and chemical relaxation of polydisperse polystyrene at 120°C in N_2.

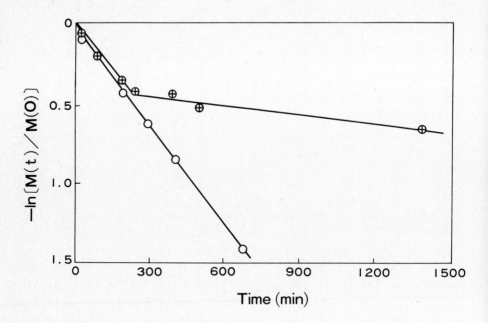

Fig. 6.3. Plots of $-\ln\{M(t)/M(0)\}$ vs. time at 120°C in N_2 for monodisperse polystyrene with dicumyl peroxide: ($\bigcirc$) 5 parts DCP, $K_1 = 2.10 \times 10^{-3}$ min^{-1}; ($\oplus$) 2 parts DCP.

There is a fast decomposition at first, followed by a slower process

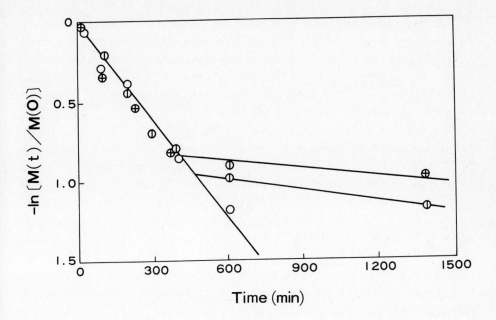

Fig. 6.4. Plots of $-\ln\{M(t)/M(0)\}$ vs. time at 120°C in N_2 for polydisperse polystyrene with dicumyl peroxide: (○) 5 parts DCP, $K_1 = 2.04 \times 10^{-3}$ min^{-1}; (①) 3 parts DCP; (⊕) 2 parts DCP.

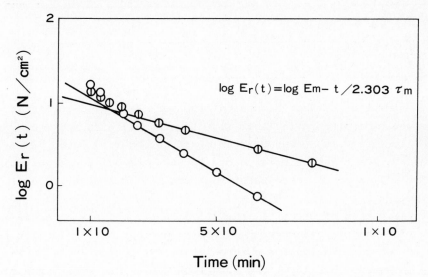

Fig. 6.5. Plots of $\log E_r(t)$ vs. time for polystyrene by Procedure X: (①) monodisperse polystyrene, $\tau_m = 1.96 \times 10^4$ min; (○) polydisperse polystyrene, $\tau_m = 4.34 \times 10^4$ min.

of decomposition for both monodisperse and polydisperse samples with smaller peroxide content.

The time required to reach the second decomposition process increases with increasing dicumyl peroxide concentration. In the case of 5 parts dicumyl peroxide, however, first-order kinetics appear to be followed over the entire range of time. In this experiment, the value of the rate constant is almost constant, independent of molecular weight distribution.

The maximum relaxation time $\tau_m(0)$ for initial molecular weight can be obtained by using the method of Procedure X (ref. 4), which is illustrated in Fig. 6.5. Assuming a discontinuous distribution of relaxation times in the rubbery-flow range, τ_m can be expressed by eq (6.9) if the time t is considerably larger than τ_m.

$$\log E_r(t) = \log E_m - t/2.303 \ \tau_m \tag{6.9}$$

The parameters, E_O and τ_e were determined from the relaxation spectrum, calculated by the Ferry and Williams second-approximation method. These results are listed in Table 6.2.

TABLE 6.2

Viscoelastic parameters for polystyrene

Polystyrene sample	E_O, dyne/cm^2	τ_m, min	τ_e, min	K_1, min^{-1}
Monodisperse	1.09×10^6	1.96×10^4	6.43×10^{-2}	2.10×10^{-3}
Polydisperse	1.09×10^6	4.34×10^4	6.43×10^{-2}	2.04×10^{-3}

The relaxation curves calculated from eqn (6.8) by substituting the viscoelastic parameters in Table 6.2 are compared with measured curve in Fig. 6.6 and 6.7. It can be seen that the measured curve deviates considerably from the calculated one for both monodisperse and polydisperse polystyrenes, except for that containing 2 parts dicumyl peroxide.

The relaxation spectrum for the polydisperse polymer tends to have a flat shape, which deviates from a box-shaped distribution function. In the calculation for both polystyrenes containing 5 parts dicumyl peroxide, eqn (6.8) was used with the value of k_1 for each sample. With 2 parts dicumyl peroxide (Fig. 6.3), k_1 is not constant through the observed time scale, so we calculated $\tau_m(t)$ directly from the M(t) values by eqn (6.6) and substituted $\tau_m(t)$ into eqn (6.8) instead of $\tau_m(0) \cdot e^{-3.4k_1 t}$. The same method of calculation was used for the samples containing 2 parts and 3 parts of dicumyl peroxide

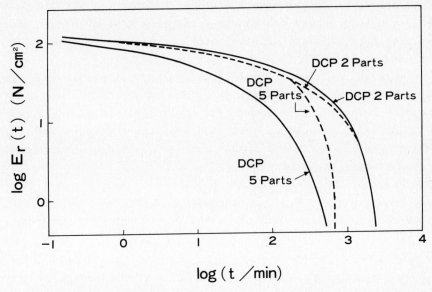

Fig. 6.6. Comparison of (---) calculated master curves with (——) experimental curves for monodisperse polystyrene: $E_O=1.09\times10^6$ dyne/cm^2, $\tau_m=1.96\times10^4$ min; $\tau_e=6.43\times10^{-2}$ min; $K_1=2.10\times10^{-3}$ min.

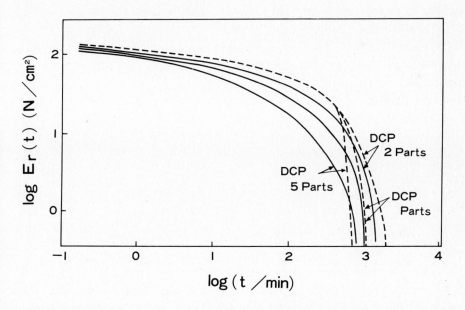

Fig. 6.7. Comparison of (---) calculated master curves with (——) experimental curves for polydisperse polystyrene. $E_O=1.09\times10^6$ dyne/cm^2, $\tau_m=4.3\times10^4$ min; $\tau_e=6.34\times10^4$ min; $K_1=2.04\times10^{-3}$ min.

for polydisperse polystyrene.

From the above experimental results, for a sample containing 5 parts dicumyl peroxide, the measured values are smaller than those calculated but good agreement was obtained for samples containing 2 parts dicumyl peroxide in both polystyrenes.

The melting point of dicumyl peroxide is 39°C, so dicumyl peroxide will be liquid and will significantly act as a plasticizer, and this is why the relaxation is greater for the sample containing 5 parts dicumyl peroxide. It can be concluded that the derived theoretical equation is valid for monodisperse samples containing low concentrations of dicumyl peroxide.

On the other hand, eqn (6.10), presented by DeBast and Gilard (ref 5) and Tobolsky (ref. 6), also adequately describes the rubbery plateau for the rubbery flow of narrow distribution polystyrenes.

$$E_r(t) = E_r(0) \cdot \exp \{- (t/\tau)^B\} \qquad (6.10)$$

In this three-parameter equation, $E_r(0)$ is the value of the rubber plateau modulus, τ is the time parameter which determines the length of the rubbery plateau, measured as the time t at which $E_r(t)$ had decayed to $E_r(0)/e$. B is the rate-of-decay parameter which influences the shape of the rubbery-flow region.

If eqn (6.10) is used as the basic equation of chemical relaxation and physical flow for linear amorphous polymers (cf. eqn (6.8)), it transforms to eqn (6.11), because parameters $E_r(0)$ and B are constant and τ is dependent on time t, though B will show a small dependence on t (ref. 7).

$$E_r(t) = E_r(0) \cdot \exp \{- (t/\tau(t))^B\} \qquad (6.11)$$

τ was found to follow relation (6.12), when expressed in seconds, for modulus curves at 115°C, according to the results of Tobolsky et al. (ref. 6).

$$\tau_{115}(t) = 1.4 \times 10^{-14} \{M(t)\}^{3.4} \qquad (6.12)$$

From eqns (6.11), (6.12), and (6.4),

$$E_r(t) = E_r(0) \cdot \exp \{- [t/1.4 \times 10^{-14} \{M(0)\}^{3.4} \cdot e^{-3.4kt}]\} \qquad (6.13)$$

When straight lines were not obtained for the relation between $-\ln \{M(t)/M(0)\}$ and t, the values of $E_r(t)$ were estimated from eqn (6.11) by using $\tau(t)$ values obtained from eqn (6.12) by substituting into the latter values of $N(t)$ in the relation of $-\ln \{M(t)/M(0)\}$ and t.

According to Tobolsky (ref. 6), there exists a series of polystyrene samples which show a relationship between B and $\bar{M}w/\bar{M}n$ ranging from B=0.608 for $\bar{M}w/\bar{M}n$=1.05 to B=0.275 for $\bar{M}w/\bar{M}n$=2.53.

By interpolating and extrapolating the values of $\bar{M}w/\bar{M}n$ for our polystyrenes into the above relation, values of B were obtained. The values of $E_r(0)$, $\tau(0)$, and B thus obtained are summarized in Table 6.3.

TABLE 6.3

Characterizing parameters in eqn (6.11) for polystyrenes

Polystyrene	$\log E_r(0)$ (dyne/cm^2)	$\tau(0)$, min	B
Monodisperse	6.96	2884	0.488
Polydisperse	7.20	301.99	0.258
Polystyrene PS-2	7.03	150.0	0.229

In order to obtain a theoretical curve from eqn (6.11), the values of $\tau(t)$ at 120°C are required. Eqn (6.14), the modified WLF equation, should be used to obtain the value of $\tau_{120}(t)$ at 120°C from that at 115°C, calculated from eqn (6.12), where T_S is the WLF temperature of polystyrene and C_1, C_2 are the universal constants.

$$\log \frac{\tau_{115}(t)}{\tau_{120}(t)} = C_2 \left\{ \frac{393-T_S}{C_1+393-T_S} - \frac{388-T_S}{C_1+388-T_S} \right\} \qquad (6.14)$$

The chemical stress-relaxation curves for monodisperse polystyrene, polydisperse polystyrene, and PS-2, the commercial polystyrene with 2 parts of dicumyl peroxide were measured at 120°C in N_2, as shown in Figs. 6.8, 6.9 and 6.10, respectively.

Three theoretical curves, one for each k value, k=0.00064 min^{-1}, 0.0021 min^{-1}, and 0.0450 min^{-1} which were arbitrarily selected, were obtained from eqn (6.13), and compared with the experimental value for PS-2 in Fig. 6.10. Good agreement between the theoretical chemical stress-relaxation curve and the experimental value is seen for both polystyrenes in Figs. 6.8 and 6.9 when compared with those in Figs. 6.6 and 6.7. This will indicate that eqn (6.11) may express the rubbery plateau region of a linear polymer more adequately than does eqn (6.8).

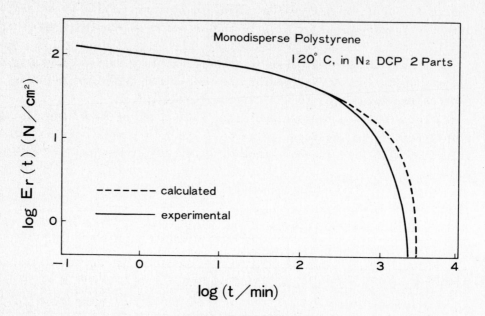

Fig. 6.8. Calculated and experimental chemical relaxation curves for monodisperse polystrene with DCP at 120°C in N_2.

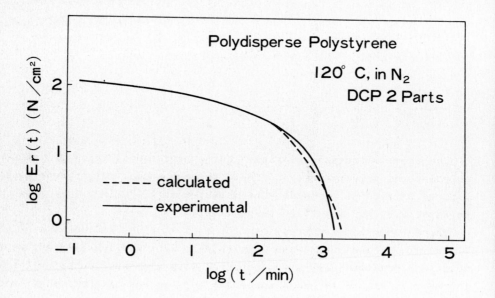

Fig. 6.9. Calculated and experimental chemical relaxation curves for polydisperse polystyrene with DCP at 120°C in N_2.

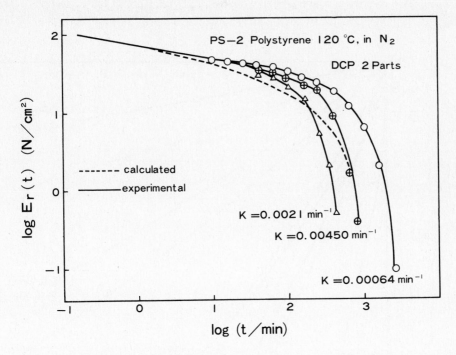

Fig. 6.10. Calculated and experimental chemical relaxation curves for PS-2 poly-styrene with DCP at 120°C in N_2.

2 Crosslinked Hydrocarbon Polymers

As shown above, chemorheological studies of uncrosslinked polymers
are rather tedious, even using a simplified theoretical treatment.
Thus, the method is hardly applicable for general practical use.
However, it is useful for crosslinked plastics undergoing chemical
reactions at high temperatures. The thermal stability of hydrocarbon
networks has recently been investigated by Shaw et al. (ref. 8) in the
temperature range 300~350°C. In this study, crosslinked low- and high-
density polyethylene (LDPE and HDPE, respectively,) were the principal
network materials, with some peroxide-cured EPT included for comparison. The stress-relaxometer was operated under vacuum or in N_2
atmosphere.

The chemical stress-relaxation curves are presented in Figs. 6.11,
6.12, and 6.13 for runs at 300, 325, and 350°C, respectively. These
temperatures cover the range of relaxation times for the networks
which is available experimentally. At temperatures much below 300°C,
relaxation is negligibly slow, while at temperatures higher than
350°C degradation is so rapid that manipulation becomes impossible.

Relaxation times for the various materials at the temperatures

176

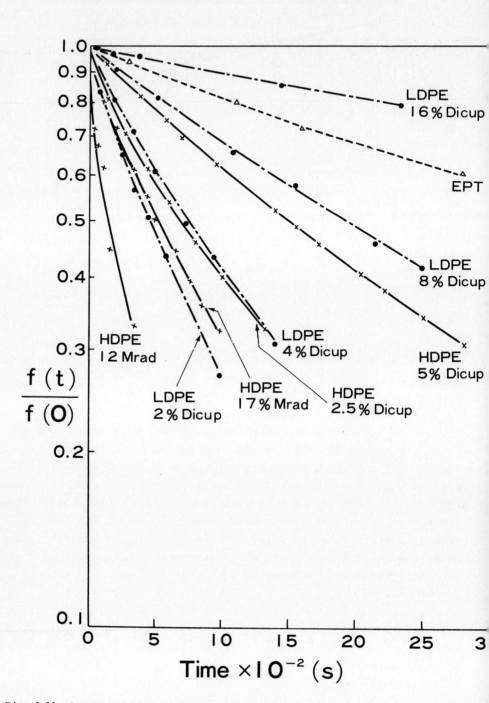

Fig. 6.11. Stress relaxation for hydrocarbon networks at 300°C in vacuum.

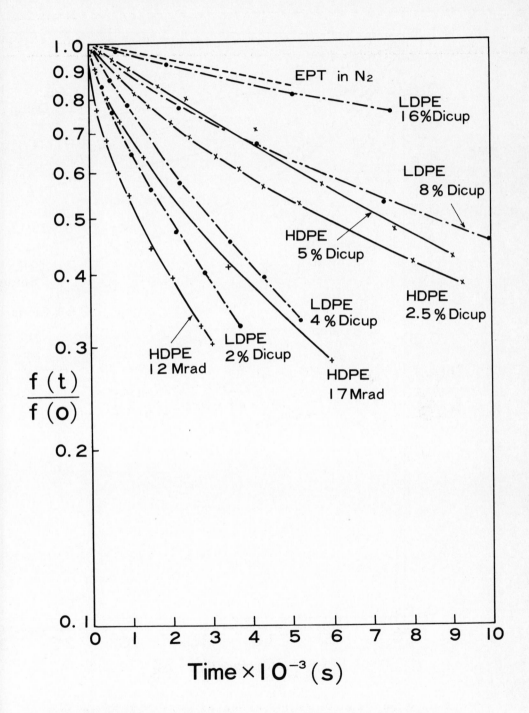

Fig. 6.12. Stress-relaxation for hydrocarbon networks at 325°C in vacuum.

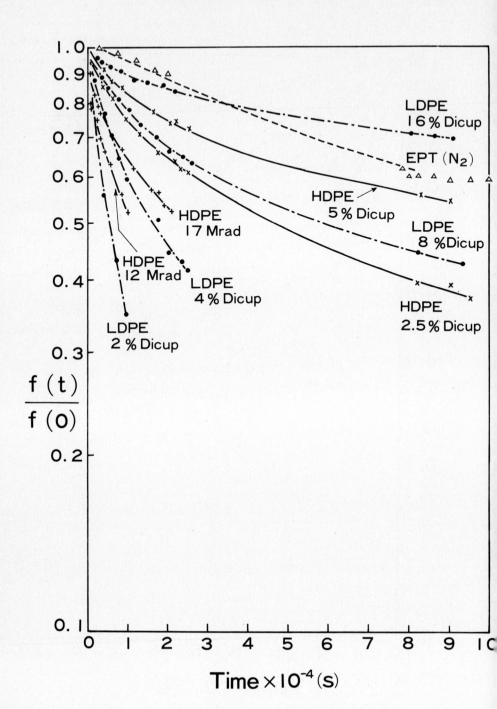

Fig. 6.13. Stress-relaxation for hydrocarbon networks at 350°C in vacuum.

studied are recorded in Table 6.4 as the times required for the
force to decay to half of its initial value ($t_{1/2}$) and 1/e of its

TABLE 6.4

Chemical stress-relaxation for hydrocarbon networks in vacuum

Polymer	Crosslinking agent	Cross-linker dose	Temp., °C	$n_e(0) \times 10^{-20}$ chains/cm^2	$t_{1/2}$, s$\times 10^{-4}$	$t_{1/e}$, s$\times 10^{-4}$
High-density polyethylene (HDPE)	Dicumyl peroxide	2.5%	300	1.38	4.7	9.8
			325	0.71	0.59	0.98
			350	0.055	0.07	0.12
		5%	300	2.25	13	-
			325	1.24	0.72	1.1
			350	0.19	0.15	0.23
	Radiation	12 Mrad	300	0.26	1.3	-
			325	0.076	0.13	0.22
			350	0.007	0.02	0.03
		17 Mrad	300	0.47	2.3	-
			325	0.22	0.24	0.42
			350	0.033	0.053	0.083
Low-density polyethylene (LDPE)	Dicumyl peroxide	2%	300	0.27	0.56	0.96
			325	0.11	0.19	0.32
			350	0.014	0.047	0.073
		4%	300	0.36	1.7	-
			325	9.23	0.29	0.47
			350	0.023	0.075	0.11
		8%	300	0.78	5.8	-
			325	0.77	0.82	1.5
			350	0.22	0.19	0.29
		16%	300	2.45	~30	-
			325	1.82	2.9	5.6
			350	1.10	0.82	1.3
EPT	Dicumyl peroxide	~1%	300	-	12	-
			325	0.23	2.0	-
			350	0.13	0.37	0.8

initial value ($t_{1/e}$), as appropriate initial network chain densities,
$n_e(0)$, computed from rubber elasticity theory, are also included in
Table 6.4.

The relaxation curves in Figs. 6.11, 6.12 and 6.13 are, in almost
every case (with EPT as the possible exception), concave upwards; that
is, the relative force decay is less rapid than exponential. This is
especially evident at the lower temperatures. Network theory requires
that first-order, irreversible cleavage of main-chain linkages, crosslinkages, or both should give a relaxation curve which is exponential
or faster than exponential. The slower-than-exponential behavior under
the conditions of these experiments means any or all of the following:
additional crosslinking reactions during the experiment, non-first-
order reactions (weak linkages), or weight loss. The latter was excluded
because correction for the weight loss (generally less than 1%) had

little effect on the shape or the placement of the curves. Additional
crosslinking or reversible reactions in these networks are generally
minor but cannot be discounted entirely. Weak linkages can most
easily account for much of the behavior of these networks, as will be
explained.

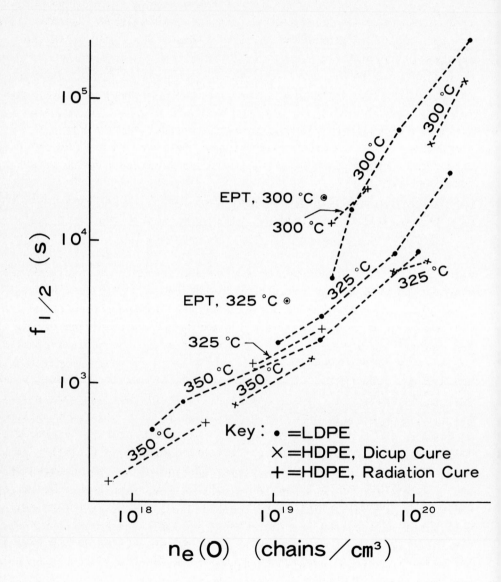

Fig. 6.14. Relaxation half-times as a function of network chain-density at various
temperatures.

In Fig. 6.14, the relaxation half-times, $t_{1/2}$, are plotted against the initial network chain-density, $n_e(0)$. Fig. 6.14 shows a definite, positive correlation between network stability and initial chain-density, which is expected for elastomers where scissions occur along the main chains of the network (See Chapter 2). That is to say, Fig. 6.14 gives the order of stability of the networks at any given cross-link density and temperature. Surprisingly, HDPE (radiation-cured) is the least stable hydrocarbon network studied, while EPT, with the greatest branching frequency, is the most stable.

Fig. 6.14 reveals some additional, less obvious generalizations which provide important clues. Taken by temperature groups, the curves in Fig. 6.14 suggest an increasing dependence of $t_{1/2}$ on $n_e(0)$ at low temperatures, a dependence which even appears to exceed first power with respect to $n_e(0)$ at 300°C. In addition, the stability of the materials at 350°C is higher than might be predicted from their behavior at 300°C and 325°C.

All of these results can be explained most realistically in terms of significantly weaker linkages present or introduced during the initial parts of the cure. The amounts need not be a large fraction of the total number of linkages, but the "weak links" present in the original hydrocarbon polymer molecules are probably not sufficiently numerous to produce all the observed effects.

Relating this premise to the experimental observations is not difficult. As mentioned earlier, the existence of the force decay, especially at low temperatures, is good evidence for weak linkages. The weak linkages cause rapid initial force decay and, as they disappear, the network stabilizes. At higher temperatures, the weak linkages break before the experiment begins: the log f(t)/f(0) versus t curves show less upward curvature, and the networks possess unexpected stability.

Shaw et al. concluded that networks based on saturated hydrocarbon polymers, cured with peroxide or radiation, decompose thermally at measurable rates (by chemical stress-relaxation) in the temperature range 300~350°C. Overall thermal stability is in the order: peroxide-cured EPT > peroxide-cured LDPE > peroxide-cured HDPE > radiation-cured HDPE. The apparent inconsistency of this result with the known greater stability of linear versus branched structures can be explained in terms of weak linkages introduced during curing.

II TEXTILES

This section deals with the application of chemorheological methods in textile research. The number of examples of such applications is

considerably less than for rubber vulcanizates, since most of the
natural and the synthetic fibers do not have crosslinks or interchain
covalent bonds. This makes it difficult to obtain information about
chemical reactions occurring in the fibers. Wool fibers, however, are
exceptional cases. Interchain cystine bonds, consisting of disulfide
linkages, play an important role in stabilizing the α-keratin structure
of native wool fibers. Fission and reformation of cystine bonds are
profoundly reflected in the mechanical properties of wool. Thus, the
major part of this section will be devoted to the chemorheology of
wool fibers.

1 Wool Fibers

The mechanical properties of wool fibers are specified by three
different regions in the longitudinal stress-strain curve (Fig. 6.15)
(ref. 9). The linear portion of the stress-strain curve up to a few

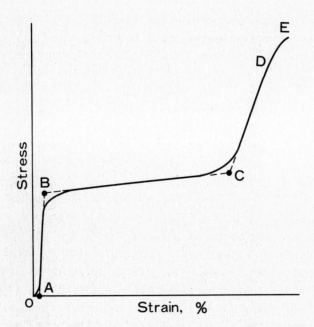

Fig. 6.15. Typical stress-relaxation curve of a keratin fiber showing the Hookean
region AB, the yield region BC, and the post-yield region CDE.

percent extension is generally referred to as the Hookean region (Fig.
6.15, AB). Beyond this region, the fiber extends much more easily
with increase of stress. This region is referred to as the yield
region (Fig. 6.15, BC). In the Hookean region, the α-keratin structure
of wool fibers remains unchanged, while in the yield region α-helices

unfold to β-pleated sheets (ref. 9). In these regions of deformation, the fiber returns to its original length after releasing in water for a sufficient time. At about 25~30% strain, the modulus of the fiber begins to increase rapidly. This region is called the post-yield region and the mechanical properties are no longer recovarable in this region, indicating that irreversible bond cleavages occur in the wool fiber.

In the native state, the α-keratin structure is stabilized by interchain cystine bonds, Coulombic interactions produced by charged groups in amino acid residues, and hydrogen bonding. Furthermore, hydrophobic interactions are particularly important in maintaining higher structures. The formation and the cleavage of these covalent or secondary bonds would contribute more or less to the changes in mechanical properties.

The first application of chemorheology to wool fibers was carried out by Katz et al. (ref. 10). Stress-relaxation was measured in water, in neutral salt solutions, and in aqueous solutions of sodium bisulfite ($NaHSO_3$) of various concentrations. The latter is known to be a reducing agent which reacts selectively with the interchain cystine bonds. The rate of stress-decay increased with increase of extension; rapid stress-decay was observed when $NaHSO_3$ was added. The rate of stress-decay in $NaHSO_3$ solutions was increased with increasing $NaHSO_3$ content

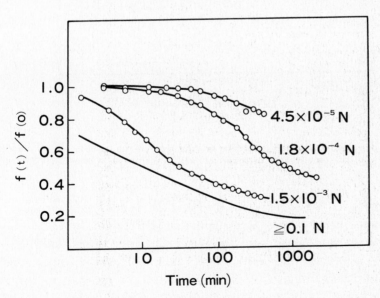

Fig. 6.16. The effect of $NaHSO_3$ concentration on the stress-relaxation of a wool fiber.

at low concentrations (Fig. 6.16) but at high concentrations the rate becomes independent of NaHSO$_3$ content. The stress-decay observed in NaHSO$_3$ solutions was ascribed to chemical reactions involving cystine linkages. Stress-relaxation in sodium chloride was also investigated. The relaxation was characterized by two-step processes.

Similar investigations were undertaken by Kubu (refs. 11, 12, 13). The effects of reducing agents such as NaHSO$_3$, thioglycolic acid, and cysteine on wool were discussed. The rates of stress-relaxation of wool fibers in the solutions of reducing agents increased markedly, while those of silk and nylon fibers, which are free from cystine bonds, were virtually unaffected by the reducing agents (Fig. 6.17). Thus, the rapid stress-relaxation observed in reducing agent solution

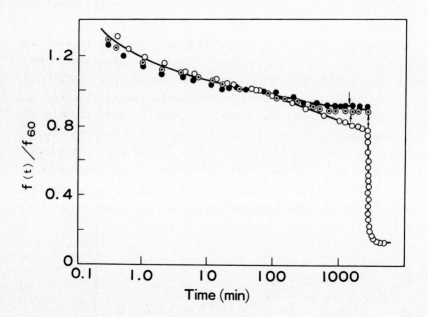

Fig. 6.17. The effect of NaHSO$_3$ addition on the stress-relaxation of (○) wool fiber, (⊙) silk and (●) nylon in water at 30°C. Arrows show the time when NaHSO$_3$ is added.

is attributable to the scission of disulfide bonds. The stress-relaxation in aqueous HCl solutions was also studied. With the additio of HCl, rapid stress-decay was observed, but the stress could be recovered by replacing the medium with distilled water. This result suggests that the stress-decay in aqueous HCl is related to the sciss of secondary bonds, especially to Coulombic interactions.

Kubu and Montgomery (ref. 12) investigated the kinetics of reactio

of cysteine solutions with cystine crosslinks as functions of concentration, pH, and temperature by means of the stress-relaxation
method. The reaction of cysteine with cystine is represented by,

$$R - S - S - R + Cys - SH \underset{k_2}{\overset{k_1}{\rightleftharpoons}} R - S - S - Cys + R - SH \qquad (6.15)$$

$$R - S - S - Cys + Cys - SH \underset{k_4}{\overset{k_3}{\rightleftharpoons}} Cys - S - S - Cys + R - SH \qquad (6.16)$$

where Cys - SH denotes cysteine. The kinetic equation for this reaction
was formulated in connection with stress-relaxation on the basis of the
following simplified assumptions.

(1) The crosslinking reaction, which is the reverse reaction in eqn
(6.15), does not contribute to the stress. This is the fundamental
assumption for analysing the continuous stress-relaxation curve of
rubber networks (see Chapter 2.1).

(2) Cysteine is added to the solution in large excess in order to
maintain its concentration virtually constant during the reaction.

(3) The difference $f(t)-f(\infty)$ is proportional to the cystine content.
Here, $f(t)$ and $f(\infty)$ are the stress at time t and the equilibrium
stress, respectively. This assumption was shown to be reasonable by
Munakata (ref. 14) in terms of the detailed chemical analysis of the
cystine content. $f(\infty)$ corresponds to the stress due to interchain
interactions other than cystine bonds.

Then, the reaction follows the pseudo first-order kinetics, as in
eqn (6.17).

$$[f(t) - f(\infty)]/f(0) = [1 - f(\infty)/f(0)] \exp(k_1[Cys-SH]_0 t) \qquad (6.17)$$

The stress-relaxation curve was fairly well explained by eqn (6.17),
as shown in Fig. 6.18. The rate of reaction decreased when the pH is
low. This was accounted for in terms of the lower dissociation constant
of the sulfhydryl groups of cysteine. The heat and entropy of activation were determined as 76 kJ/mol and 8.8 e.u., respectively.

Another important application of chemorheology to wool fibers is
the investigation of setting process. When a wool fiber is held
extended in hot water or steam for fairly long periods, and then
released, the fiber retains some of its extended length. The fiber
is said to be "set". In order to clarify the setting mechanism, a
number of chemorheological studies were attempted in water or steam

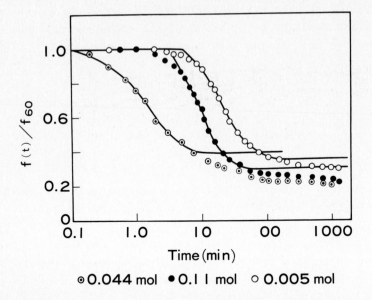

Fig. 6.18. Chemical stress-relaxation of wool fiber in aqueous cysteine solutions
Temperature: 50°C, Elongation: 15%, pH: 6.5.

at elevated temperatures.

The rate of stress-decay was studied in relation to the second-
order transition temperature (Tg) of α-keratin (ref. 15). Tg is considered to be the lower limit for setting treatments. Below Tg, the
stress-decay was ascribed to the reversible transformation of native
keratin. On the other hand, the higher rate of stress-decay above Tg
was considered to be associated with non-reversible bond cleavages.

More detailed studies on the stress-relaxation behavior in hot
water were carried out by Weigmann et al. (ref. 16). According to
their results, the stress-relaxation of wool fibers is best regarded
as a 2-stage process. The first stage is characterized by rapid
stress-decay with a small activation energy. This process was though
to be the relaxation of secondary bonds. The rate of stress-decay
in the second stage is slower than that of the first stage. The
stress-decay curves in phosphate buffer at pH 7.0 are shown in Fig.
6.19. It can be seen that the second stage becomes prominent at high
temperatures. The relaxation modulus at 10 s after elongation is
plotted against temperature in Fig. 6.20. The effect of the addition
of N-ethylmaleimide (NEMI), which is the well-known sulfhydryl block
reagent, is also shown in this figure. The pronounced stress-decay i
the second stage was observed above the transition temperature (70°C

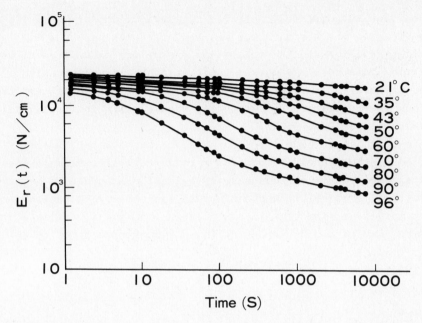

Fig. 6.19. The plot of relaxation modulus versus time for wool fibers in phosphate buffer of pH 7.0.

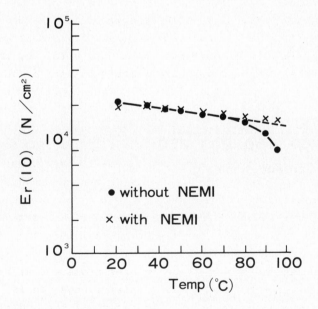

Fig. 6.20. Relationship between relaxation modulus of a wool fiber and temperature. The relaxation modulus $E_r(10)$ is determined at 10 s after elongation. pH: 7.0, Elongation: 20% (500%/min.).

This reaction is completely retarded by the addition of NEMI. The stress-decay in the second stage was well approximated by a simple Maxwellian relation and the activation energy was calculated to be 96 kJ/mole. This result suggests that the controlling mechanism of the stress-decay in this stage is the sulfhydryl-sulfide interchange reaction. The transition temperature decreases with increasing sciss of cystine bonds (Fig. 6.21).

The sulfhydryl-sulfide interchange reaction in water is also indicated by the dynamic mechanical method (ref. 17). A small strain was superposed on the various rate of extension in order to determin dynamic Young's moduli in Hookean, yield and post-yield regions. Dynamic moduli at different rates of extension were approximately superposed on one another by using the WLF shift factor up to 60°C.

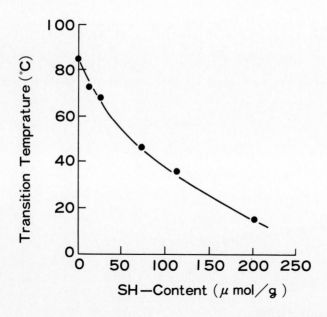

Fig. 6.21. Relationship between the transition temperature of a wool fiber and sulfhydryl content.

Above 60°C, however, the data did not fit the WLF equation, indicati the onset of scission and exchange reactions involving cystine bonds

More fundamental studies on the mechanism of the second stage of stress-relaxation in water were carried out by Miyauchi et al. (ref. 18). A model fiber of wool, disulfide-crosslinked polycaprolactam (DSPC) fiber, was prepared and the stress-relaxation behavior of thi fiber was compared with that of the wool fiber. The stress-relaxatio

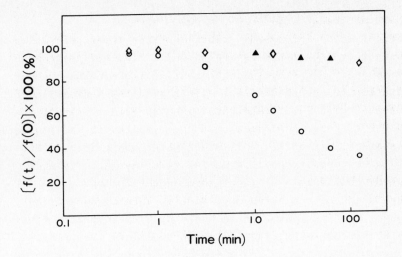

Fig. 6.22. Stress-relaxation curves for DSPC fibers in water at 90°C: pre-immersion time, 10 min.; elongation, 100%; (○), untreated; (▲), treated with 0.08M KCN at 50°C for 4 h; (◇), treated with 2,4-dinitrofluorobenzene.

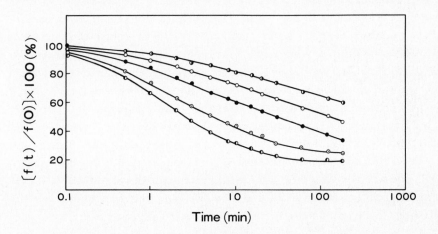

Fig. 6.23. Stress-relaxation curves for wool fibers of different lanthionine contents in water at 90°C: pre-immersion, 10 min; elongation, 25%; estimated lanthionine contents (μmol/g), (◐) 0, (☉) 135, (●) 297, (○) 342, (◑) 393.

behavior of DSPC and wool fibers pretreated with potassium cyanide was also discussed. It has been well established that the cystine cross-links of wool can be quantitatively converted to lanthionine (mono-sulfide) crosslinks by the action of potassium cyanide. As shown in Fig. 6.22, the stress-decay for an untreated DSPC fiber in hot water was quite rapid. On the other hand, the stress-decay of DSPC fibers pretreated with KCN or a sulfhydryl blocking reagent (2,4-dinitro-fluorobenzene) was retarded markedly. This confirms that the controlling mechanism of the stress-decay for DSPC fibers in hot water is the sulfhydryl-sulfide interchange reaction. The stress-relaxation curves of KCN-treated wool fibers are shown in Fig. 6.23. Although the rate of stress-decay decreased with increasing lanthionine content, the stress-relaxation of the treated wool fiber with the highest lanthionine content proceeded with a moderate rate, in contrast to that of the KCN-treated DSPC fibers. According to these results, it was suggested that the stress-relaxation of wool in hot water was primarily due to cleavage of the secondary bonds in native keratin, which is much promoted by the sulfhydryl-sulfide interchange reaction.

The contribution of the sulfhydryl-sulfide interchange to the second stage of stress-relaxation in water is also supported by the suppression of stress-relaxation in low pH solutions (ref. 19). As described previously, most of the sulfhydryl groups would be undissociated in these conditions.

Recently, several studies were undertaken to clarify the contribution of secondary bonds to stress-relaxation. The basic role of hydrophobic interactions in stabilizing native α-keratin structure was first pointed out by Zahn (ref. 20). It was found that the Young's moduli of wool fibers were much lower in a water-alcohol mixture than in distilled water The moduli decreased with increasing alcohol content up to a minimum, beyond which the fiber became stabilized. These results were explained in terms of the breakage of hydrophobic interactions in the less polar alcoholic medium. The stress-relaxation of wool fibers in aqueous alcohols were investigated by Nemoto et al. (ref. 21). The rate of stress-decay increased with the chain length of the alcohol and the maximum relaxation was observed in 45~60% alcohol solutions (Fig. 6.24 and Fig. 6.25). The stabilization of wool fibers at high alcohol concentrations was ascribed to the higher stability of Coulombic interactions and hydrogen bonds in the less polar solvents (ref. 20). The chemical stress-relaxation was also investigated in aqueous alcohols using sodium bisulfite as reducing agent (Fig. 6.26). The rate of stress-decay was higher in alcoholic solutions than in water, which may be attributable to the increased accessibility of cystine bonds on breaking up the hydrophobic regions (ref. 21).

The role of Coulombic interaction in mechanical cycling was investigated by Feughelman (ref. 22). In the Hookean and the yield regions, it was established that the strain is recoverable after releasing. This result is regarded as evidence that none of the irreversible bond cleavages occurs in these regions. The mechanical

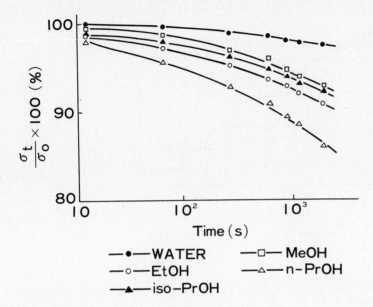

Fig. 6.24. Stress-relaxation curves for a wool fiber in 45 vol.% alcohol solution (20% extension, at 22±1°C). σ_o and σ_t are the stress at time zero and time t, respectively.

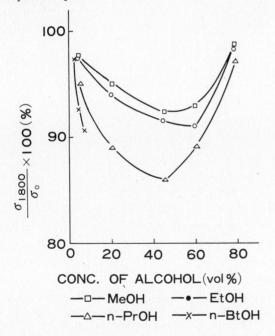

Fig. 6.25. Relation between the relative stress for wool fibers at 1800 s after immersion in aqueous alcohol solutions and alcohol concentration. σ_{1800} represents the stress at 1800 s after the immersion.

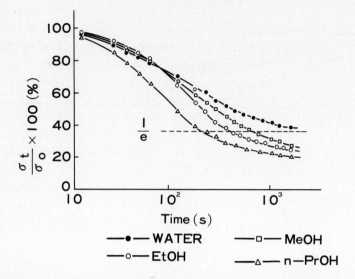

Fig. 6.26. Stress-relaxation curves for wool fibers in 60 vol.% aqueous alcohol solution containing 0.1 mol of NaHSO$_3$.

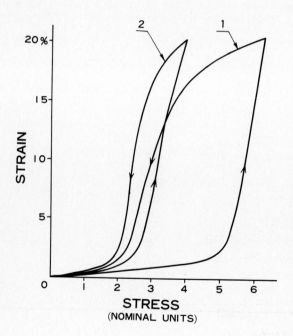

Fig. 6.27. The hysteresis loops for a wool fiber in distilled water (Curve 1) and in aqueous HCl at pH 1 (Curve 2). The rate of extension and retraction is 0.1%/min.

cycling in water, however, was by no means reversible in the visco-

elastic sense. Mechanical hysteresis loops of wool fibers in water
or in aqueous HCl are shown in Fig. 6.27. It can be seen that the size
of the hysteresis loop is considerably reduced in aqueous HCl. The
elimination of negatively charged groups, such as the carboxylic acid
groups of the glutamic and aspartic acid residues, could be responsible for the reduction of the hysteresis loop. Thus, the larger
hysteresis loops and the slower recovery of the stress during the
retraction cycle in distilled water are probably due to Coulombic
interactions being broken during extension and reformed with great
difficulty. It was concluded that the Coulombic interactions play an
important role in stabilizing the α-helix structure of wool fibers.

In summary, chemorheological methods have been widely applied to
elucidate the nature of various interchain interactions in wool
keratin fibers.

2 Applications to the Other Fibers

The applications of chemorheology to the fibers other than wool are
limited because of the absence of interchain crosslinks. Among these,
the estimation of accessibility of fibers to chemical reagents is
most useful from a practical viewpoint.

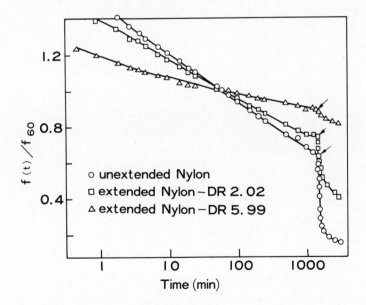

Fig. 6.28. Stress-relaxation of extended and unextended nylon fibers in water.
Arrows show the time when 1N HCl was added.

Lemiszka et al. (ref. 23) investigated the stress-relaxation behavior of various synthetic and natural fibers in distilled water, with and without chemical reagents. The stress of Nylon fell rapidly with the addition of HCl (Fig. 6.28), while the stress-relaxation curves of polyester fibers did not change with the addition of HCl. With cellulose fibers, rapid stress-decay was observed in 0.1N HCl. This is due to cleavage of the glucoside linkages in the molecular chains of cellulose. Lemiszka estimated the relative accessibilities of cellulose fibers to HCl in terms of the extrapolated intercepts of the linear parts of the $f(t)/f(0)$ curves. The values thus obtained were consistent with those obtained by other methods, indicating the applicability of stress-relaxation to the estimation of accessibility. Now, it is well known that cellulose fibers contain some distribution of the ordered regions, ranging from the totally amorphous to the completely crystalline. The boundary between the crystalline and amorphous regions is made obscure by the presence of intermediate regions, in which the degree of order varies continuously from completely disordered to completely ordered. The chemical and physical properties of cellulose fibers are much influenced by the distribution of ordered regions.

Empirically, eqn (6.18) is used to describe the decay of stress for cellulose fibers immersed in aqueous HCl (ref. 23).

$$f(t)/f(0) = \sum_{n=1}^{s} A_n e^{-k_n t} \qquad (6.18)$$

where A_n and k_n are parameters which specify the fraction and the relaxation frequency of an order of type n. The physical meaning of equation (6.18) was made clear by Canter (ref. 24). According to his treatment, the system (fiber network) was divided into sub-systems. The time-dependence of the number of these sub-systems was obtained by assuming that the bond-cutting obeys Markovian statistics. The calculated relative stress was of the same form as in eqn (6.18), indicating that the exponential stress-decay was the natural consequence of the Markovian cutting process. The validity of this theory, however, is difficult to justify, since equations similar to (6.18) can easily be derived from first-order random scission of the network chains (see Chapter 2.1).

A quantitative measure of the effect of water on stress-relaxation was developed for various fibers by Whitney et al. (ref. 25). The breakdown of the secondary bonds was considered to be the principal

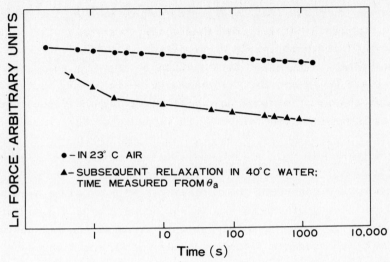

Fig. 6.29. Effect of hot water on strained cellulose diacetate. (●) in air at 23°C; (▲) subsequent relaxation in water at 40°C. Time measured from θa, when the sample was immersed.

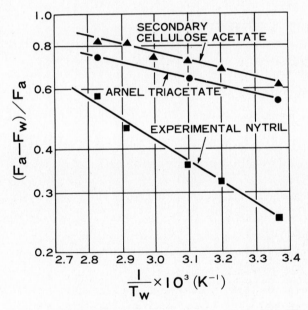

Fig. 6.30. Magnitude of the water effect on various fibers as a function of water temperature.

cause of the stress-decay. After the fiber had relaxed in air for time θa, it was submerged in water, controlled at a constant tempera-ture. The resulting plot of logarithmic force versus time showed two distinct linear portions (Fig. 6.29). It was proved that the first portion did not result from diffusion effects, but was ascribed to the breakdown of secondary bonds by hydration. The effect of water was best described in terms of the function $(F_a-F_w)/F_a$, as shown in Fig. 6.30, where F_a is the force in air at time θa and F_w is the force at the end of the water effect. It can be seen that the dependence of the effect on water temperature T_w is represented by,

$$\ln\left[(F_a - F_w)/F_a\right] = K_1 - (K_2/T_w) \tag{6.19}$$

where K_1 and K_2 are constant. The function $(F_a-F_w)/F_a$ gives a measure of the hydrophilicity of the fiber. It was found that the behavior of a fiber in water could be interpreted in terms of the redistribution of forces among the various secondary bonds within the fiber.

The stress-relaxation for Dinitrile A fibers in aqueous dimethyl formamide (DMF) solutions was studied by McMickle et al. (ref. 26). The stress-relaxation produced by the addition of 50% DMF-water solution is shown in Fig. 6.31. The force first increased rapidly and

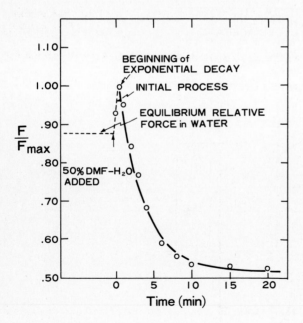

Fig. 6.31. Stress-relaxation of a Dinitrile A fiber in 50% DMF - water at 59°C. The solid curve is calculated by assuming exponential stress-decay.

then decreased until an equilibrium value was attained. The stress-decay was closely approximated by an exponential curve. The effects

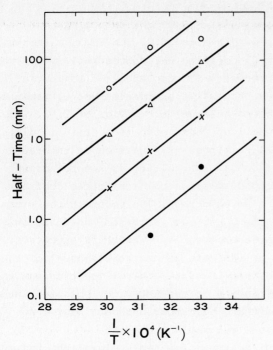

Fig. 6.32. The plot of observed half-life, $t_{1/2}$, as a function of the reciprocal absolute temperature for various DMF concentrations. (○), 20% DMF; (△), 35% DMF; (X), 50% DMF; (●), 65% DMF.

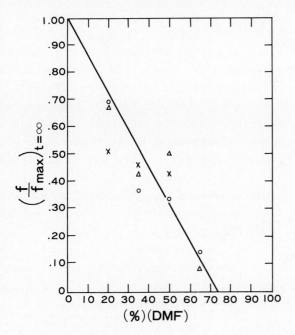

Fig. 6.33. The plot of the extent of relaxation $(f/f_{max})_{t=\infty}$ as a function of DMF concentration. (○), at 30°C; (△), at 45°C; (X), at 59°C.

of temperature and concentration of DMF on the rate of stress-decay
of Dinitrile A fibers are shown in Figs. 6.32 and 6.33. Although the
rate seemed to increase with increasing DMF concentration, the
activation energy for the process was independent of concentration.
It was emphasized that three successive processes must be taken into
account to interpret the stress-relaxation behavior of fibers in
terms of the kinetics of reaction. These processes are: (1) diffusion
of the reagent to the crosslink sites, (2) reaction of the reagent
with these sites to produce bond-scission, and (3) rearrangement of
the polymer molecule after bond-scission to a preferred configuration
under the condition of the imposed strain. The second process is
generally much faster than the other two processes, so it is only
necessary to compare the rates of the first and third processes. The
above-described dependence of the rate of the stress-decay on the
DMF concentration indicated that the rate-controlling step was the
first process, i.e., the diffusion of the DMF-water solution into
the fiber. It was confirmed that chain rearrangement played no
significant role in the stress-relaxation behavior.

In view of the above applications of stress-relaxation, the method
is especially of use to estimate the accessibility of an aqueous mediu
to hydrophilic fibers.

III PAINTS AND ADHESIVES

Most of the polymers utilized as paints and adhesives are comprised
of thermosetting resins. The structure of these resins is characterize
by a three-dimensional network, which serves to improve mechanical
properties, and chemical and thermal stability. At high temperatures,
these resins show rubber elasticity, when the moduli of the resins
become proportional to the number of crosslink sites. Hence, the usual
chemorheological methods are easily applied in relation to the thermal
stability of these resins.

The stress-relaxation for thermosetting acrylic resins in hot water
was extensively investigated by Hakozaki et al. (ref. 27) using epoxy
and melamine resins as curing agents. Curing by epoxy resin was
carried out at 140 and 160°C. After pretreatment in boiling water for
90 min, the stress-relaxation was measured (Fig. 6.34). The stress-
relaxation curves were analyzed using Procedure X (ref. 4), as shown i
Fig. 6.35. The relative stress could be expressed by eqn (6.20).

$$f(t)/f(0) = Ae^{-k_1 t} + Be^{-k_2 t} + Ce^{-k_3 t} \qquad (6.20)$$

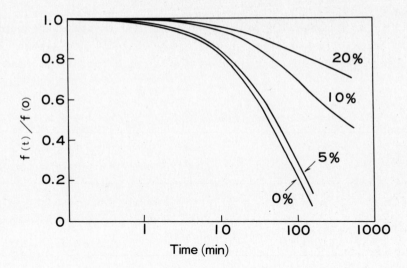

Fig. 6.34. Effect of the content of epoxy resin on the stress-relaxation of acrylic resins in water at 98°C. Acrylic resins are cured at 140°C.

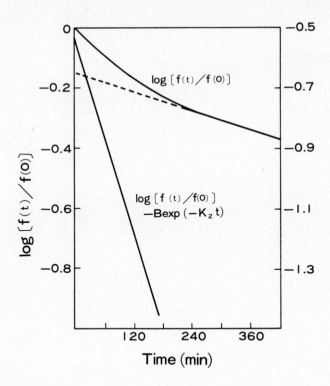

Fig. 6.35. Resolution of the stress-relaxation curve for acrylic resin, containing 10% epoxy resin, by Prodecure X.

where   A + B + C = 1.

The sets of rate constants for stress-decay and the pre-exponential factors for acrylic resins with various epoxy contents are listed in Table 6.5. The stress-relaxation curves for acrylic resins with and

TABLE 6.5

Stress-relaxation constants for epoxy (Epon 1001)-cured acrylic resins at 140°C

Epoxy resin content (wt.%)	A	$k_1 \times 10^2$ (min⁻¹)	τ_1 (min)	B	$k_2 \times 10^3$ (min⁻¹)	τ_2 (min)	C	$k_3 \times 10^4$ (min⁻¹)	τ_3 (min)
0	1	1.63	61						
5	1	1.44	69						
10	0.284	1.22	82	0.708	1.18	848			
20	0.180	1.18	88						
	0.023	8.82	11				0.795	2.81	3580

without 5% epoxy resin were expressed by single exponential term, while those containing 10% and 20% epoxy resin were expressed by two and

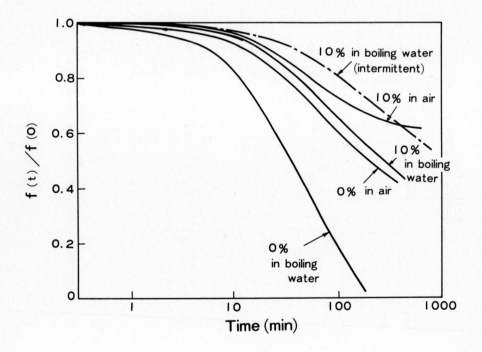

Fig. 6.36. Stress-relaxation curves for epoxy-cured acrylic resins, either in air or in water at 98°C. The weight percent of epoxy resin is indicated in the figure.

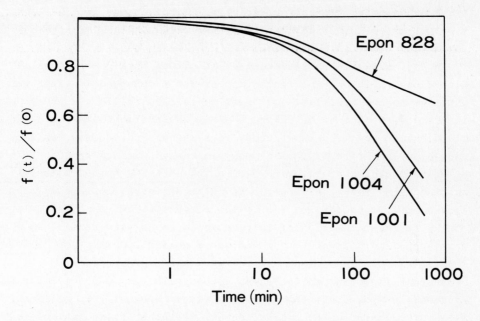

Fig. 6.37. Effect of the molecular weight of epoxy resin on the stress-relaxation of epoxy-cured acrylic resins in boiling water. Epoxy content is 10%. The number-average molecular weights of Epon 828, Epon 1001 and Epon 1004 are 390, 1000 and 1850, respectively.

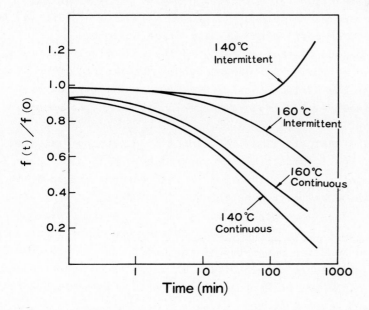

Fig. 6.38. Continuous and intermittent stress-relaxation curves for melamine-cured acrylic resin.

three exponential terms, respectively. Contributions from various types of crosslink structure may be reflected in the discrete set of relaxation times in eqn (6.20).

In Fig. 6.36, the stress-relaxation curves in air are compared with those in water. The faster stress-decay observed in water was ascribed to the hydrolysis of crosslink sites, while the stress-decay in air was thought to be due to thermal degradation alone. The effect of molecular weight between crosslinks was also discussed. As shown in Fig. 6.37, the rate of stress-decay increased with increasing length of epoxy chain, that is, with decreasing network chain-density. It was suggested that the slower rate of stress-decay for highly cross-linked acrylic resins was due to low accessibility to water.

The stress-relaxation of self-crosslinked acrylic resins was also carried out in hot water. The energy of activation was about 130 kJ/mol indicating that the scission of network chains took place.

Another important curing agent for acrylic resins is melamine resin. As shown in Fig. 6.38, the stress-relaxation behavior in boiling water of melamine-cured resins is much different from that observed for epoxy-cured resins. The fast stress-decay in the initial stages may be attributed to the elution of unreacted melamine resin. On the other hand, it was confirmed by means of the intermittent method that the crosslinking reaction for unreacted melamine proceeds in the later stages. These results suggest that the curing of acrylic resins by melamine should be carried out at high temperatures.

Another application of chemorheology to thermosetting polymers is the investigation of curing processes. Dynamic mechanical methods can be used to follow the curing process, but certain limitations exist because of the liquid-solid transition taking place in the course of the reaction.

Torsional braid analysis (TBA), developed by Gillham et al. (ref. 28) circumvented the difficulties encountered by the liquid-solid transition. The basic concept involved in this elegant method is to support the resin on an inert substrate such as glass fiber braid and to determine the dynamic mechanical properties of the composite by the damped free oscillation method. Even in the rubbery or flow regions the inertial mass would be supported by the braid, and so the measurements could be extended to the condition in which the current torsional pendulum method could not be applied. A schematic diagram of TBA is given in Fig. 6.39 (ref. 29).

The curing processes of various thermosetting resins were monitored by TBA (ref. 30). As shown in Fig. 6.40, the method gives a well-

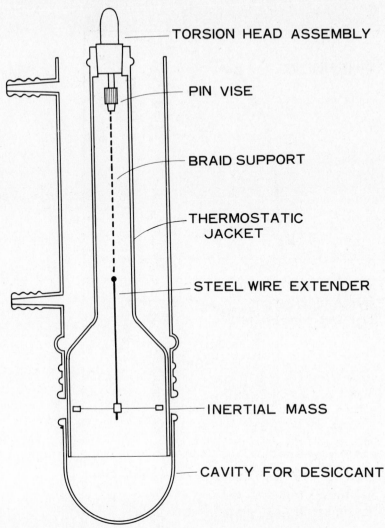

TORSION HEAD ASSEMBLY

PIN VISE

BRAID SUPPORT

THERMOSTATIC JACKET

STEEL WIRE EXTENDER

INERTIAL MASS

CAVITY FOR DESICCANT

Fig. 6.39. Schematic diagram of torsional braid analysis.

defined measure of the degree of curing.

Changes in mechanical properties obtained by TBA during the thermosetting of hexa-allylmelamine were compared with the chemical conversion of allyl groups (Fig. 6.41) (ref. 31). It is seen that the increase of rigidity and the mechanical damping peak correspond well to the progress of the crosslinking reaction. The method could be applied to the studies on the thermal softening behavior, and pyrolytic and oxidative degradation, as well as on curing behavior (ref. 30).

A modification of TBA was recently proposed by Naganuma et al.

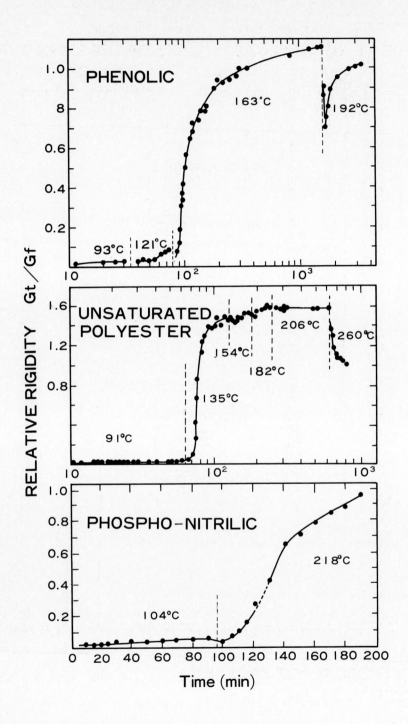

Fig. 6.40. Curing behavior of thermosetting resins.

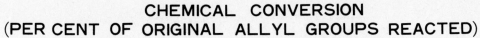

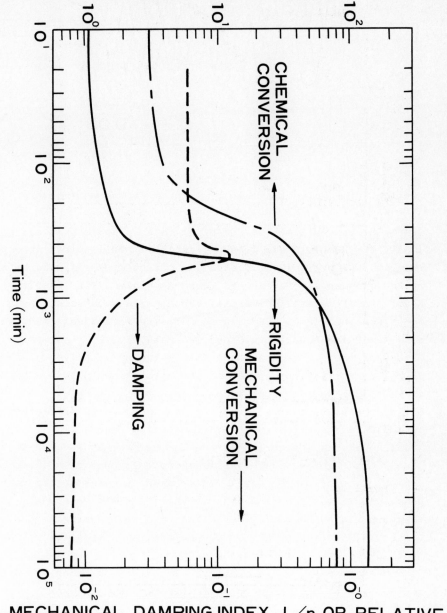

CHEMICAL CONVERSION
(PER CENT OF ORIGINAL ALLYL GROUPS REACTED)

Fig. 6.41. Thermosetting of hexa-allylmelamine at 130°C.

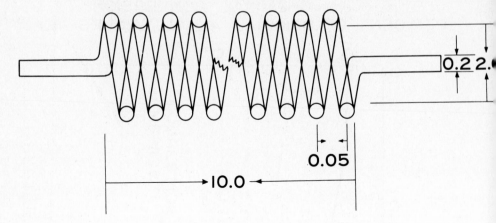

Fig. 6.42. Dimension of the standard spring (in millimeter) for DSA.

(ref. 32). In this method, referred to as dynamic spring analysis
(DSA), the sample resin is supported by a small spiral spring
instead of a braid (Fig. 6.42). The curing process could be monitored
by the changes in the dynamic Young's modulus. The crosslinking
reactions of several commercial cyanoacrylate adhesives were followed
by this method (Fig. 6.43) (ref. 33). The method was found to be
more sensitive than the current TBA method.

One of the disadvantages of TBA is the qualitative nature of the
method. Exact interpretation of the data is not possible because
the absolute value of the modulus cannot be obtained, nevertheless,
it is especially useful as a practical method for estimating the
optimum condition for curing thermosetting resins.

Although the monitoring of the curing process as a whole cannot
be achieved by the usual dynamic mechanical method, studies on post-
curing behavior can be easily undertaken by means of the torsion
pendulum. It was reported that at the gel point only 70% of the
reaction has been completed and that the post-curing beyond the gel
point has a marked effect on the mechanical properties of the resin
(ref. 34). Thus, careful dynamic mechanical studies on the post-
curing stage seem to be important. The study on amine-cured epoxy
resins is probably the most typical one.

Much attention has been paid to the characterization of the dynamic
mechanical loss peaks of amine-cured epoxy resins (refs. 35, 36).
The secondary relaxation, termed γ, has been observed to depend on
the state of cure (ref. 35). Arridge et al. examined in detail the
dependence of the γ-relaxation on the state of cure (ref. 37). As

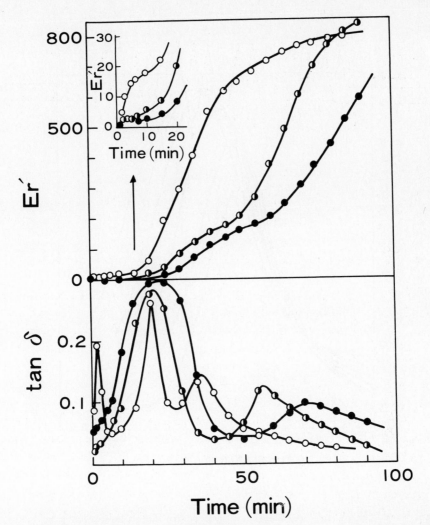

Fig. 6.43. Crosslinking reactions of commercial cyanoacrylate adhesives monitored by DSA at 20°C.

shown in Fig. 6.44, the peak value of tan δ and the temperature of tan δ_{max} increase with increase of the state of cure. The shear modulus was also plotted as a function of the state of cure (Fig. 6.45) (ref. 37). It is seen that the shear modulus below the γ-transition increases with increasing crosslinking, while above the

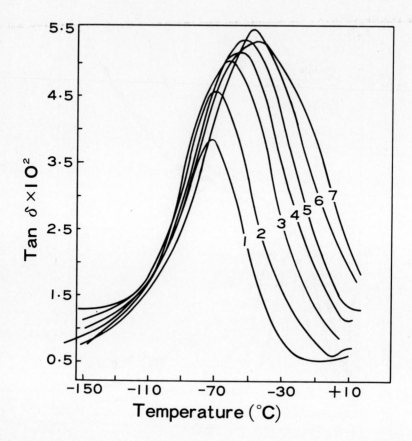

Fig. 6.44. Change in γ-relaxation with cure for an amine-cured epoxy resin.

transition the reverse is true. These phenomena were interpreted in terms of the increase of segmental mobility arising from the disappearance of epoxy groups. It was proposed that the ratio of the shear modulus above and below the γ-transition may be used as a quantitative measure of cure in the post-curing region (Fig. 6.46).

The foregoing experiments on thermosetting resins show that the mechanical method can be utilized in the quantitative estimation of the curing process, if the relation between mechanical behavior and the chemical conversion is established. Owing to the simplicity and the sensitivity of the measurements, the mechanical method has become increasingly important in the adhesive and paint industries.

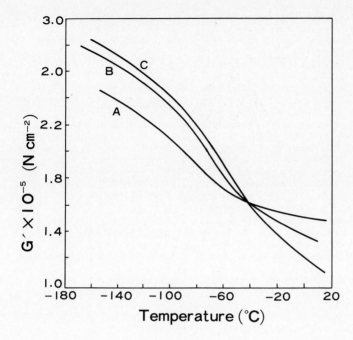

Fig. 6.45. Shear modulus-temperature relation for varing cures. Cure temp. for samples A, B and C are 19°C, 84°C and 159°C, respectively.

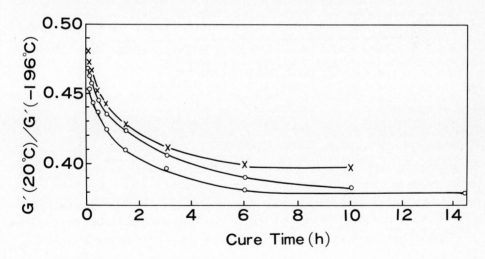

Fig. 6.46. Modulus ratio as a function of cure time for three different cure temperatures: (X), 120.8°C; (●), 132.8°C; (○), 145.6°C.

REFERENCES

1 A.V. Tobolsky, Properties and Structure of Polymers, Wiley, New York, 1960, p.124.

2 K. Murakami, S. Tamura, and H. Nakanishi, J. Polym. Sci., Polym. Chem. Ed., 11(1973)313; K. Murakami and H. Nakanishi, ibid., 14(1976)265.
3 T. Kusano and K. Murakami, Bull. Chem. Res. Inst. Non-aqueous Solutions (Tohoku Univ.), 20(1970)199.
4 A.V. Tobolsky and K. Murakami, J. Polym. Sci., 40(1959)443.
5 J. DeBast and P. Gilard, Phys. Chem. Glasses, 4(1963)117.
6 W.F. Knoff, I.L. Hopkins, and A.V. Tobolsky, Macromolecules, 4(1971)750.
7 K. Murakami and H. Nakanishi, J. Polym. Sci., Polym. Chem. Ed., 14(1976)489.
8 M.T. Shaw and A.V. Tobolsky, J. Polym. Sci., Part A-1, 9(1971)1937.
9 E.G. Bendit and M. Feughelman, in Encyclopedia of Polym. Sci. and Tech. Vol.8, Wiley, New York, 1968, p.20.
10 S.M. Katz and A.V. Tobolsky, Text. Res. J., 20(1950)87.
11 E.T. Kubu, ibid., 22(1952)765.
12 E.T. Kubu and D.J. Montgomery, ibid., 22(1952)778.
13 E.T. Kubu, F. Frei and D.J. Montgomery, ibid., 24(1954)659.
14 H. Munakata and Y. Niinami, Sen-i Gakkaishi (Journal of the Society of Fiber Science and Technology, Japan), 19(1963)64.
15 B.J. Rigby, Text. Res. J., 29(1959)926.
16 H.D. Weigmann, L. Rebenfeld and C. Dansizer, ibid., 36(1966)535.
17 P. Mason, Kolloid-Z. Z. Polym., 218(1967)46.
18 S. Miyauchi, K. Kajiyama, M. Sakamoto and H. Tonami, Appl. Polym. Symposia, No.18(1971)925.
19 M. Feughelman and M.S. Robinson, Text. Res. J., 39(1969)196.
20 H. Zahn, Kolloid-Z. Z. Polym., 197(1964)14; H. Zahn and G. Blankenburg, Text. Res. J., 34(1964)176.
21 Y. Nemoto, Y. Terada and S. Terada, Sen-i Gakkaishi (Journal of the Society of Fiber Science and Technology, Japan), 27(1971)473.
22 M. Feughelman, J. Macromol. Sci. - Phys., B7(1973)569.
23 T. Lemiszka and J.C. Whitwell, Text. Res. J., 25(1955)947.
24 N.H. Canter, ibid., 36(1966)220.
25 C.K. Whitney and R.L. Hamilton, J. Polym. Sci., A2(1964)3577.
26 R.H. McMickle and E.T. Kubu, J. Appl. Phys., 26(1955)832.
27 J. Hakozaki, S. Hashimoto and E. Higashimura, Zairyo, 16(1967)484; idem. ibid., 16(1967)487.
28 A.F. Lewis and J.K. Gillham, J. Appl. Polym. Sci., 6(1962)422.
29 A.F. Lewis and J.K. Gillham, ibid., 7(1963)685.
30 J.K. Gillham and A.F. Lewis, ibid., 7(1963)2293.
31 J.K. Gillham and J.C. Petropoulos, ibid., 9(1965)2189.
32 S. Naganuma, T. Sakurai, Y. Takahashi and S. Takahashi, Kobunshi Kagaku, 29 (1972)105.
33 S. Naganuma, T. Sakurai, Y. Takahashi and S. Takahashi, ibid., 29(1972)519.
34 Y. Tanaka and H. Kakiuchi, J. Appl. Polym. Sci., 7(1963)1951.
35 D.E. Kline, J. Polym. Sci., 47(1960)237.
36 A.S. Kenyon and L.E. Nielsen, J. Macromol. Sci. - Chem., A3(1969)275.
37 R.G.C. Arridge and J.H. Speake, Polym., 13(1972)443.

216

Styrene-butadiene copolymers (SBR), 1, 49, 60
Sulfhydryl blocking reagents, 189
Sulfhydryl groups, of cysteine, 185
Sulfhydryl-sulfide inter-change reaction, 188, 190
Swelling, equilibrium degree of, 46
Synthetic rubber, 15
Tensile stress, 58
Tetrafunctional crosslinkages, 118
Tetrahedral model of network structure, 98
Textiles, 181
Thermal degradation, 202
Thermal scission of cross-links, 113
Thermosetting resins, 198
Thioglycolic acid, 184
Time lag, 39
Torsional braid analysis (TBA), 202
Torsion pendulum, 206
Transient experiments, 35, 39
γ-Transition, 207, 208
Trisulfide linkages, 62
Two network theory, 141, 143
Two stage process, of stress relaxation, 186
Ultra-Violet irradiation, 146
Uniform network chain-length, 7, 12
Van de Graaf source, 75
Vulcanization, 156
Weak linkages, 179
WLF shift factor, 188
Water, effect of, on stress relaxation, 196

Yield point, 128
Yield region, 182, 190
Young's moduli, of wool fibers, 190